Circuits and Fields

Circuits and Fields

A First Course

P. AARON

and

W.N. TABERNER

PRENTICE HALL

London   New York   Toronto   Sydney   Tokyo   Singapore
Madrid   Mexico City   Munich

First published 1995 by
Prentice Hall International (UK) Limited
Campus 400, Maylands Avenue
Hemel Hempstead
Hertfordshire, HP2 7EZ
A division of
Simon & Schuster International Group

Typeset in 10/12pt Times by
Mathematical Composition Setters Ltd, Salisbury, Wiltshire

Printed and bound in Great Britain by
T. J. Press Ltd, Padstow

Library of Congress Cataloging-in-Publication Data

Available from the publisher

British Library Cataloguing in Publication Data

A catalogue record for this book is available from
the British Library

ISBN 0-13-341819-7

2 3 4 5   99 98 97 96 95

Contents

Preface

This book is intended as an introductory course in the topics of Electromagnetic Fields and Electrical Circuits for Electrical Engineers at first-degree level. The book may also be of interest to other disciplines such as physics.

The book confines itself to material that can be expected to be tackled satisfactorily in the first year of an undergraduate course and deliberately excludes more advanced work that may be undertaken in second- and third-year courses. The treatment is fundamental and aims to give a solid base from which students can tackle problems of a varied nature, not only of the type covered in the problems and worked examples of this text. This ability is one of the distinctions between a technician and a true degree student.

The mathematics needed for the book has been surveyed in Chapter 2. This chapter is deliberately brief and to the point to allow students to get quickly to the main topics of the book. Where necessary enhancement of the mathematics is given in the text and appendices. Although the mathematics needed to understand the book has been restricted in depth, vectors, integration, and complex numbers have not been excluded. These topics are considered essential for a proper treatment of the subjects and to provide a secure base for further study.

The circuits and fields are combined in a single volume since the authors believe that they are the two fundamental subjects that all students taking an Electrical course should study. It is useful to have the fundamentals of a discipline set out clearly and concisely in one book. Parts 2 and 3, although complementary, may be read independently. The book could be the basis of a single study module.

Chapter 13 is included to emphasize to students that circuits and fields are descriptions of the same phenomena and are closely linked, and to emphasize the modelling aspect. A detailed comparison of the two subjects involves mathematics beyond the scope of this introductory text.

References have not been included in the text or at the end of chapters. With such a fundamental book many excellent texts can be used to supplement the material presented. A restricted list of these books is provided in the bibliography.

PART ONE

Introduction

CHAPTER ONE

Units and dimensions

Units and dimensions are of fundamental importance. In addition they form a short reference to the major features of a subject. The dimension describes the physical characteristic of a quantity, and the unit a standard by which the quantity can be measured.

1.1 The SI system

The international system of units is known as the *SI system* and electrical work is based on five units, namely *kilogram* (mass), *metre* (length), *second* (time), *ampere* (current), and *kelvin* (temperature). From these fundamental units all other units may be derived, e.g. coulombs = amperes × seconds. Two other fundamental units, namely those for luminous intensity, the candela, and amount of substance, the mole, are not used here.

The *dimension* of these units is an indication of their origin. Thus the dimension of the unit kilogram is M which is shorthand for Mass. The dimensions of any derived quantity may be denoted in terms of the dimensions of the four basic units, i.e. M (mass), L (length), T (time) and I (current).

A list of the most important units used in mechanical and electrical systems is shown in Table 1.1 In this table the units, their symbols, and their dimensions are quoted.

To develop the dimensions of Table 1.1 it is necessary to know the relationships between the various quantities and equate them until the necessary dimensions are established. If the dimensions for power, P, and current, I, are known, then the dimension for voltage may be found from the equation $P = IV$ by noting that

The dimension of $V = \text{dim of } P / \text{dim of } I = ML^2T^{-3}/I = ML^2T^{-3}I^{-1}$

The units listed above are defined in the following chapters. It is important when doing problems that a student should always work with

3

Table 1.1 Units and dimensions

Quantity	Unit	Symbol	Dimension
Mass	kilogram	kg	M
Length	metre	m	L
Time	second	s	T
Current	ampere	A	I
Temperature	kelvin	K	–
Velocity	metres/s	m/s	LT^{-1}
Acceleration	metres/s/s	m/s^2	LT^{-2}
Force	newton	N	MLT^{-2}
Energy, work	joule	J	ML^2T^{-2}
Power	watt	W	ML^2T^{-3}
Charge	coulomb	C	IT
Voltage	volt	V	$ML^2T^{-3}I^{-1}$
Electric field	volts/m	V/m	$MLT^{-3}I^{-1}$
Electric flux density	coulombs/m^2	C/m^2	MTL^{-2}
Capacitance	farad	F	$M^{-1}L^{-2}T^4I^2$
Permittivity	farads/m	F/m	$M^{-1}L^{-3}T^4I^2$
Resistance	ohm	Ω	$ML^2T^{-3}I^{-2}$
Conductance	siemens	S	$M^{-1}L^{-2}T^3I^2$
Inductance	henry	H	$ML^2T^{-2}I^{-2}$
Permeability	henries/m	H/m	$MLT^{-2}I^{-2}$
Magnetic flux	weber	Wb	$ML^2T^{-2}I^{-1}$
Magnetic flux density	tesla	T	$MT^{-2}I^{-1}$

the units as specified above to ensure the integrity of the units of the answer.

1.2 Dimensional analysis

The 'dimensional accuracy' of answers may be used to check that gross mistakes have not been made in an analysis. For example, suppose a problem has been carried out and the answer is expected to be *energy*. The quantities involved in the answer to the problem are (capacitance)(voltage)^2 or CV^2. The dimensions of CV^2 are $(M^{-1}L^{-2}T^4I^2)(ML^2T^{-3}I^{-1})^2 = ML^2T^{-2}$, and this is the dimension of energy.

1.3 Unit multiples and submultiples

The units in the table are often too big or small for specific problems. Therefore standard prefixes are used as shown in Table 1.2.

Table 1.2 Prefixes

Prefix	Abbreviation	Magnitude
Tera	T	10^{12}
Giga	G	10^{9}
Mega	M	10^{6}
Kilo	k	10^{3}
Milli	m	10^{-3}
Micro	μ	10^{-6}
Nano	n	10^{-9}
Pico	p	10^{-12}
Femto	f	10^{-15}

CHAPTER TWO

Introductory mathematics

This chapter contains a résumé of the minimum mathematical material that is needed to provide a foundation for the topics in the book. It is not intended to take the place of a suitable mathematics textbook, but to highlight the relevant mathematics and give useful results which may help the student. Some material in this chapter is enhanced in appendices to supplement understanding at this or a later stage. It is hoped that many students will have been exposed to most of the ideas and results quoted here, and the chapter will provide mainly revision and a 'quick look-up' facility. Others may find new material here; for them it will be necessary to use a suitable mathematics textbook, providing a full and detailed coverage of the new material.

2.1 Complex numbers

Complex numbers are used in order to deal effectively with problems in alternating current work. The operator j, which is an imaginary quantity, is a fundamental concept in complex number theory. A complex number is one which has both real and imaginary parts. Graphical representation of complex numbers is made in the *complex plane* where real quantities are abscissae and imaginary quantities are ordinates.

2.1.1 The operator 'j'

All quantities in the complex plane are referred to the horizontal *x*-axis and the vertical *y*-axis, and to differentiate between the two axes we preface all *y*-axis components with the symbol 'j'. In effect 'j' means move anticlockwise through 90°. It is also common to refer to the *x*-axis as the *real* axis and the *y*-axis as the *imaginary* axis.

Since j causes a 90° anticlockwise rotation, j^2 will cause a rotation of 180° and this is equivalent to reversing the direction. Therefore $j^2 = -1$, and $j = (-1)^{1/2}$; j is therefore an *imaginary* quantity, and a number prefaced by j is known as an imaginary number.

A simple example of a complex number is $(2 + 5j)$.

2.1.2 The Argand diagram

The basic features of complex numbers are illustrated by the Argand diagram, which is a diagram in the complex plane (Figure 2.1).

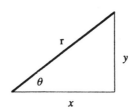

Figure 2.1 An Argand diagram

The diagram demonstrates that the quantity r may be represented by the expression

$$r = x + jy \tag{2.1}$$

and when $|r|$ is the magnitude of r, then the resolved parts of r along the real and imaginary axes are $x = |r| \cos \theta$ and $y = |r| \sin \theta$. From the diagram it is clear that $|r| = (x^2 + y^2)^{1/2}$.

An important identity relating trigonometric and exponential statements is

$$e^{j\theta} = \cos \theta + j \sin \theta \tag{2.2}$$

Multiplying equation (2.2) by $|r|$ we obtain an expanded version of equation (2.1):

$$r = |r|e^{j\theta} = |r| \cos \theta + j|r| \sin \theta \tag{2.3}$$

The quantities $|r|$, x, y, and θ are real and r is complex. It may also be noted that the magnitude of $e^{j\theta}$ is always unity and $e^{j\theta}$ merely specifies direction.

2.2 Manipulation of complex quantities

Complex numbers can be manipulated easily and thus provide a means of dealing with problems involving angles as well as magnitudes analytically rather than graphically. It is only necessary to remember that real and imaginary quantities must be treated *separately*. Let us consider the two complex numbers $x = a + jb$ and $y = c + jd$ and work out the results of addition, subtraction, multiplication, and the complex conjugate.

2.2.1 Addition and subtraction

Using the two complex numbers

$$x + y = (a + c) + j(b + d)$$

and

$$x - y = (a - c) + j(b - d)$$

the additions and subtractions are normal, but the real and imaginary parts are treated separately.

2.2.2 Multiplication

$$xy = (a + jb)(c + jd) = ac - bd + j(bc + ad)$$

The brackets are multiplied out in the normal way, j^2 being replaced by -1; real and imaginary parts are again treated separately in the addition process.

2.2.3 Magnitude and angle

The modulus gives the magnitude of the complex number. It is denoted by the symbol | | and is obtained by squaring the real parts and the imaginary parts, adding them together and forming the square root. This gives the length of the hypotenuse as shown in Figure 2.2. The angle θ is the direction in which the resultant magnitude acts. It is important to realize that $|a + jb| = |a - jb|$, but the resultant magnitude acts at a different angle.

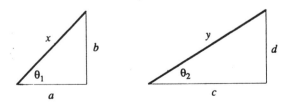

Figure 2.2 Components of two complex numbers x and y

Therefore we have the following expressions assuming x is acting at angle θ_1 and y at angle θ_2:

$$|x| \quad = (a^2 + b^2)^{1/2} \qquad\qquad \text{angle } \theta_1 = \tan^{-1}(b/a)$$
$$|y| \quad = (c^2 + d^2)^{1/2} \qquad\qquad \text{angle } \theta_2 = \tan^{-1}(d/c)$$
$$|xy| \quad = [(a^2 + b^2)(c^2 + d^2)]^{1/2} \qquad \text{angle } \theta = \theta_1 + \theta_2$$
$$|x/y| \quad = (a^2 + b^2)^{1/2}/(c^2 + d^2)^{1/2} \qquad \text{angle } \theta = \theta_1 - \theta_2$$

2.2.4 Complex conjugate

The complex conjugate of a complex number is found by changing the sign of the imaginary term. Therefore the complex conjugates of the complex numbers $(a + jb)$ and $(c + jd)$ are $(a - jb)$ and $(c - jd)$.

If a number is multiplied by its complex conjugate then the imaginary terms cancel and we are left with the modulus of the number squared:

$$(a + jb)(a - jb) = a^2 + b^2$$

2.3 Simultaneous linear equations

When solving linear algebraic equations in several variables it is often necessary to solve several equations at the same time. Such a set of equations which are interrelated are called *simultaneous equations* and an example is given below. To solve for n variables it is a requirement that n equations be available. For two variables x and y we need two equations. Consider the two simple simultaneous equations $x + 4y = 12$ and $2x + 12y = 32$.

The equations are easily solved by eliminating one of the variables by subtraction. Therefore if we multiply the first equation by 2 and subtract from the second equation then $4y = 8$, giving $y = 2$. Substituting in either of the original equations will then give $x = 4$. More equations and variables require the above technique to be repeated.

The values for x and y may also be found by organizing the equations using determinants and applying Cramer's rule. Consider the two equations below

$$ax + by = p$$
$$cx + dy = q$$

where a, b, c, d, p, and q are coefficients or constants and x and y are variables.

The determinant for the equations is

$$\Delta = \begin{vmatrix} a & b \\ c & d \end{vmatrix} = ad - bc$$

Then by Cramer's rule

$$x = \det \begin{vmatrix} p & b \\ q & d \end{vmatrix} / \Delta$$

and

$$y = \det \begin{vmatrix} a & p \\ c & q \end{vmatrix} / \Delta$$

For the numerical example $\Delta = 12 - 8 \qquad = 4$
therefore $\qquad x = (144 - 128)/4 = 4$
and $\qquad y = (32 - 24)/4 \quad = 2$

which agree with the results already found. For this problem Cramer's method is complicated compared with the simple subtraction technique. However, for large numbers of equations it provides a routine method that may be easily done by computer.

2.4 Graphs

It is often necessary to obtain information from data given in graphical form. Consider an independent variable x and a dependent variable y. An example graph of these variables is shown in Figure 2.3. Section OA is a straight line with equation $y = mx + c$. Section AB is a straight line with equation $y = $ constant. Let us assume that the equation describing section BC is not known.

The rate of change of y with respect to x is given by the derivative dy/dx. This is represented on the graph as the *slope* of the graph at any point. For Figure 2.3 the slope is positive and constant along OA, zero along AB, and negative and variable over BC.

The definite integral $\int_0^{x'} y \, dx$ represents the area under the graph from O to x'. For a graph such as Figure 2.3 the value of each section may be

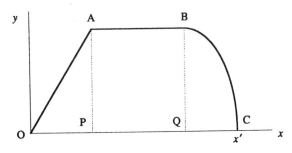

Figure 2.3 A graph having three distinct sections OA, AB, and BC

found separately since there is a different function for sections OA, AB, and BC.

If the graph can be represented as a known function $y = f(x)$ and the function is simple, then the slope and area may be found by normal calculus techniques as shown in the example below. For section BC where the equation is not known, the slope and area may be found by measurements on the graph. The slope at a point requires an estimate of the slope of the tangent to the curve at the point. The area may be found by counting squares under the graph outline. An example follows.

————— **Example 2.1** —————————————————————————————————

Find the slope and area under the graph for Figure 2.4. The portion OA follows a square law.

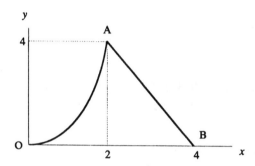

Figure 2.4

The problem must be treated in two parts, namely sections OA and AB. In this problem both sections are represented by simple functions.

The slope

The function for OA is

$$y = x^2$$

Therefore

$$dy/dx = 2x$$

Putting in values for x gives the slope at all points between O and A. The function for AB is

$$y = mx + c$$

The straight line traced back cuts the y-axis at a value 8. The slope of the line is -2. The equation of the line is therefore

$$y = -2x + 8$$

and

$$dy/dx = -2$$

as expected.

The area

The area under OA is

$$\int_0^2 x^2 \, dx = [x^3/3]_0^2 = 2.67$$

The area under AB is

$$\int_2^4 (-2x + 8) dx = [-x^2 + 8x]_2^4 = 4$$

The value 4 is expected since it is the area of a triangle height 4, base 2. These answers may be corroborated by measuring the slope and area from the graph. This is left as an exercise for the reader.

2.5 Scalars and vectors

Fields exist in three-dimensional space and may point in any direction. It is therefore essential to describe them in terms of vectors, even in an introductory course, since only then can the groundwork for future study be established.

The vector symbolism ensures correct descriptions of field quantities and allows equations to be written in a *concise* style. The use of vector algebra only arises when doing problems. Even then it is often simpler to work using sketches and treating the problem trigonometrically.

For the above reason the details of vector algebra are relegated to Appendix 4, and in the present section the minimum material necessary to appreciate equations written in a vector format is presented.

2.5.1 Definitions

Scalars are quantities that have *magnitude* only. Examples of scalars are energy, power, pressure within a fluid, temperature. Scalars are written in normal type.

Vectors are quantities with both *magnitude* and *direction*. Examples of vectors are velocity, force, electric field. Vectors are indicated by using **bold type**. As an example the electric field E is written as **E**.

Problems where direction is important are often difficult to visualize. Although direction can be represented by using vectors there is still a need for good diagrams.

It is good practice to make a sketch for all problems – the sketch improves understanding, and often suggests a means for tackling problems.

2.5.2 Coordinate systems

In this text the Cartesian system, a set of three mutually orthogonal axes x, y, and z (Figure 2.5) will be used and will be perfectly adequate for the problems presented. Further information on coordinate systems may be found in Appendix 4 and more advanced texts on electromagnetism.

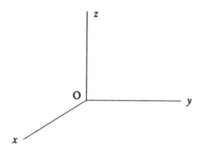

Figure 2.5   Cartesian axes

The axes are at right angles to each other and by convention are arranged such that a right-handed screw turning from x to y advances along the z-axis.

2.5.3 Unit vectors

A unit vector is a vector specifying a particular direction, and having unit magnitude. This is a convenient way of separating the magnitude and direction of a vector. It is useful to be able to specify a unit vector along an arbitrary direction in space, as for example along a vector direction denoted by the symbol **r**. It is necessary to show clearly that the magnitude is unity, and this is done by using a 'carot' symbol over the vector, i.e. $\hat{r}$. This technique is used in the presentation of the vector equations in this text.

Therefore if in Figure 2.6 we have a vector force **F** of magnitude 10 along the line AB that has direction $\hat{r}$ in space, then the vector may be represented by $\mathbf{F} = 10\hat{r}$.

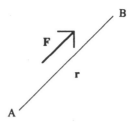

Figure 2.6   A vector **F** acting along line AB with direction **r**

Special symbols exist for unit vectors along the Cartesian axes. They are $\mathbf{a}_x$, $\mathbf{a}_y$, and $\mathbf{a}_z$ for unit vectors along the x-, y-, and z-axes respectively. A force of magnitude 15 along the y-axis can be represented by $\mathbf{F} = 15\mathbf{a}_y$.

2.6 Scalar and vector products

Also known as *dot* and *cross* products these functions replace 'normal' multiplication when using vector quantities. They are invoked to provide a method for dealing with the physical problems encountered in situations where directions are important.

2.6.1 The scalar product

Consider two vectors **A** and **B** that make an angle with each other of θ, the angle being in the same plane as the vectors. Then the scalar product is defined as

$$F = \mathbf{A} \cdot \mathbf{B} \tag{2.4}$$

F is a scalar and has magnitude $AB \cos \theta$, where θ is the *smaller* angle between the two vectors.

This involves *resolution* since one vector is resolved along the direction of the other before 'multiplication'. The result of the action is a *scalar quantity*. The dot between **A** and **B** is not just a full stop, it is a *new vector operator* and must always be marked in very clearly.

2.6.2 The vector product

Consider two vectors **A** and **B** that make an angle with each other of θ, the angle being in the same plane as the vectors (Figure 2.7). Then the vector product is defined as

$$\mathbf{F} = \mathbf{A} \times \mathbf{B} \tag{2.5}$$

F is a vector and has the magnitude $AB \sin \theta$ where θ is the smaller angle

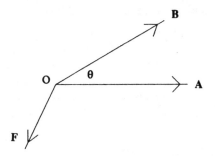

Figure 2.7 The direction of **F** for the vector product **F = A × B**, is out of the plane of the paper

between **A** and **B**. The direction of **F** is perpendicular to both **A** and **B**, and such that a clockwise screw motion from **A** to **B** advances along **F**.

The cross is not a normal multiplication sign but a *new vector operator* and must be shown clearly. Note that the operator is a concise expression that avoids constant repetition of a relatively lengthy description in words.

___ **Example 2.2** _____

Two vectors **A** and **B** with magnitudes 5 and 9 act as shown in Figures 2.8 and 2.9. Find the scalar and vector products in each case.

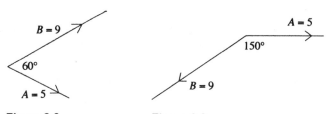

Figure 2.8   **Figure 2.9**

For Figure 2.8 $A \cdot B = 45 \cos 60° = 22.5$
For Figure 2.9 $A \cdot B = 45 \cos 150° = -38.97$
In each case the smaller angle is taken. Note how a negative sign appears since **A** when resolved is effectively in the opposite direction to **B**.
For Figure 2.8 $|A \times B| = 45 \sin 60° = 38.97$
and **A × B** points out of the page.
For Figure 2.9 $|A \times B| = 45 \sin 150° = 22.5$
and **A × B** points into the page.

2.7 Incremental vectors and integrals

Line and surface vectors are needed to provide a means of specifying incremental values to be used in *line* and *surface integrals*. Such integrals are a common feature of electromagnetic vector equations. They are needed because in electromagnetism values of quantities summed up along lines, and over surfaces and volumes, are a common occurrence owing to the three-dimensional nature of the subject.

2.7.1 Incremental length vector

A very small length d*l* along a contour C at point P (Figure 2.10) when specified as a vector **dl** gives the direction of the contour at P and has a small magnitude d*l*. The total length of the contour may be given by the following expressions:

$$\text{Length of } open \text{ contour} = \int_C dl \quad \text{and} \quad \text{length of } closed \text{ contour} = \oint_C dl$$

When the contour is closed, for example a circle or a closed polygon, a small circle is superimposed on the integral sign. If the contour C is a circle of radius r then the integral would have a value $2\pi r$.

The vector **dl** is used with the scalar product to resolve other vectors along the contour. If we have a force **F** then **F** · **dl** has a value F d*l* cos θ where θ is the angle between **F** and **dl**. This effectively resolves F along the contour.

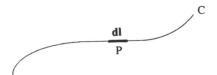

Figure 2.10   An elemental vector length dl on contour C

2.7.2 Incremental surface vector

A very small area of surface d*s* at P on a surface S when specified as a vector **ds** (Figure 2.11) has a direction *normal* to the surface at P and *out* of the surface, and a small magnitude d*s*.

The total area of the surface is given by the expressions

$$\text{Area of } open \text{ surface} = \int_S ds \quad \text{and} \quad \text{area of } closed \text{ surface} = \oint_S ds$$

When the surface is closed, for example a sphere or a cube, a circle is placed on the integral sign. If the surface is a sphere of radius r then the integral would have a value of $4\pi r^2$.

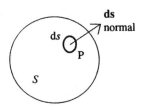

Figure 2.11 An elemental vector area **ds** on surface S

The vector **ds** is used with a scalar product to resolve other vectors along a normal to the surface. If we have a vector **F** then **F** · **ds** has a value $Fds \cos \theta$ where θ is the angle between **F** and **ds**. This resolves F along the normal to the surface at ds.

Line and surface integrals are used with scalar and vector products to write electromagnetic equations in a concise manner. Examples of the way in which line and surface integrals can be used is left until later sections dealing with specific electromagnetic problems.

Circuits

Circuit analysis and circuit theory are titles commonly used to describe techniques for analysing electrical and electronic circuits. Precisely defined component parameters, laws, and theorems are used to analyse accurately circuit diagrams. These diagrams are an attempt to 'model' the real world circuit using the ideal components of circuit theory. They are essentially approximations although often very accurate ones.

In this book the emphasis is to cover these analysis techniques fundamentally so that the reader will be able to deal with any circuit likely to be encountered.

To help the reader achieve this aim Chapter 3 provides clear and concise definitions of the basic terms and introduces the fundamental laws. This is followed by a detailed coverage of circuit analysis techniques using only resistances and d.c. sources. Experience shows that this is the best way to develop a firm understanding without the additional complications of a.c. circuits. The necessary mathematics for dealing with a.c. circuits is introduced and it is then comparatively simple to apply all the laws, theorems, and analysis techniques to any circuit, a.c or d.c.

Circuit analysis techniques do not rely on a detailed knowledge of the physical nature of the components and this is emphasized in this part of the book. The interesting total engineering problem is in the modelling process and simple examples of this are introduced where appropriate.

CHAPTER THREE

Components and basic laws

This chapter defines the basic electrical quantities and components; conventions for polarities are also introduced. This is to achieve a clearly defined base for subsequent work. It is assumed that most readers will have met these terms before and will have given some thought to the nature of electric charge and to the idea of electric conduction, leading to the concept of current as the movement of charge. However, circuit analysis does not depend on a particular view of the physical nature of the electrical properties of matter. It is important to be clear and precise in thought and to use only clearly defined terms.

The two fundamental electrical laws, due to Ohm and Kirchhoff, are stated and used.

3.1 Definitions of basic quantities

Charge

Electrical charge is normally denoted by the letter Q and the practical unit of charge is the coulomb (C).

Circuit analysis is more concerned with current than charge and hence a more detailed treatment of charge is left to Part 3. However, it must be recognized that both positive and negative charges can exist.

Current

An electric current is a movement of charge in a particular direction. An electric current, I, is defined as the rate of flow of charge.

Thus

$$I = dQ/dt \tag{3.1}$$

21

By convention a positive current in a particular direction occurs when positive charges move in that direction. (This is a straightforward, convenient, and universally used convention to which the reader will soon become accustomed. Initial confusion sometimes arises as theory suggests that current in a conductor is a physical movement of electrons and electrons are negatively charged. The movement of positive charges in one direction is equivalent to a movement, at the same rate, of negative charges in the other direction. Simply use the convention that a positive current is a flow of positive charge.)

The practical unit of current is the ampere (A). The shortened form amp is frequently used.

1 ampere is a flow of charge of 1 coulomb per second

Potential or voltage

The potential difference or voltage difference, V, between two points is defined as the energy gained, or lost, by a unit positive charge in moving between the points.

The practical unit is the volt (V) which has the dimensions of joules/coulomb. In electrical circuits voltage is *always* used in the sense of a voltage difference between two points. Any mention of a voltage at a point always implies the voltage difference between that point and some understood reference point. This might be ground (also referred to as earth) or a connection common to several components.

It is important in analysing circuits to be very precise about the sign of a voltage difference. Although + and − symbols are sometimes used in circuit diagrams it is convenient to use an arrow symbol when labelling the voltage between two points. The arrow in Figure 3.1 denotes the voltage difference $V_A - V_B$.

It should be noted that

10 V is more positive than 5 V
−5 V is more positive than −10 V

Figure 3.1   Labelling the polarity of a voltage

Power

Power is the rate of doing work or the rate of expenditure of energy.

The energy W expended in moving charge Q through a potential difference V is given by

$$W = QV \qquad (3.2)$$

Thus power,

$$P = dW/dt = V\, dQ/dt = VI$$

A power of 1 watt (W) is equivalent to a current of 1 A flowing through a voltage drop of 1 V.

Note that, in general, voltage and current are time-varying quantities. Throughout the book lower case letters are used to indicate the instantaneous values of time-varying quantities. The most fundamental definition of electric power is

$$p = iv \qquad (3.3)$$

The instantaneous value of power, p, seldom has any practical significance. The *average* value of power over a given time is a more useful quantity. If p is plotted as a function of time, its mean value over any time interval is obtained by calculating the area under the curve for that time interval and dividing by the time interval. Integration may be required to calculate the area beneath the curve. This will arise again later when considering alternating currents and voltages. When voltage and current are both constant this leads to the expression VI obtained earlier. The reader is advised against the not uncommon mistake of multiplying the average value of the current by the average value of the voltage. This is not the same as the average of the iv product and hence will not result in a correct value for average power.

3.2 Basic components

3.2.1 Resistance

In a given material free electrons move under the influence of an externally applied voltage; this is a current flow. Collisions with atoms in the material cause a loss of energy and impede the flow of current.

The resistance, R, of the component is defined by the relationship

$$R = V/I \qquad (3.4)$$

where V is the voltage across the resistance and I is the current through it. This is known as Ohm's law and the unit of resistance is the ohm (Ω).

It is important to have a clear sign convention to deal with the direction of the current and the polarity of the voltage (Figure 3.2). For a resistance the voltage is positive at the end at which current enters.

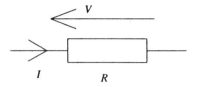

Figure 3.2   Sign convention for voltage and current

A resistance can dissipate energy but not store it.

The rate of dissipation of energy, or power loss, is given by

$$P = VI = RI^2 = V^2/R \qquad (3.5)$$

Power does not have sign other than that a negative value for power dissipated implies a power source and a negative power delivered is actually a power absorbed. In performing an analysis it may not be known whether a particular part of a circuit is acting as a net supplier of power or is absorbing power. A practical example of this is a car battery. When in normal use it is acting as a source. When it is connected to a charger it is absorbing power.

If R is a constant then the V/I characteristic of the component is a straight line as shown in Figure 3.3.

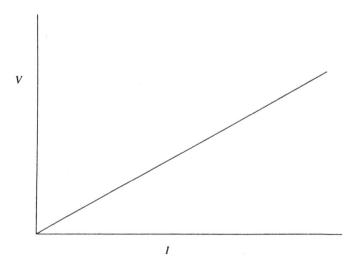

Figure 3.3   A linear resistance

The component is then said to be *linear*; that is, the *response (V)* is directly proportional to the *stimulus (I)*.

It will be noticed here that it is just as valid to speak of a voltage drop being caused by a current flow as it is to speak of a current flow caused by an applied voltage.

There are many examples of non-linear components – magnetic devices, filament lamps, diodes, transistors.

Analysis techniques exist for non-linear problems but it is often possible to use linear methods in what is known as *small signal analysis*. Consider the diagram of Figure 3.4.

Let the voltage and current be V_a and I_a for a non-linear device with the characteristic shown. The point 'a' corresponding to these values is called the *operating point*. Any small change in voltage Δv, and corresponding change in current Δi, will move the operating point along the characteristic. The effective resistance is $\Delta v/\Delta i$. This corresponds to the slope of the tangent to the characteristic at 'a' and is known as the *slope* or *incremental* resistance. Note that this is quite different to the ratio V_a/I_a, which is the slope of a line drawn from 'a' to the origin. Over a small range of values in V and I about point 'a' the characteristic and the tangent are approximately the same. Thus the component is presumed to be approximately linear over a given small working range. The resistance to be used in such calculations must be the slope resistance.

The circuit analysis in this book will assume linear components.

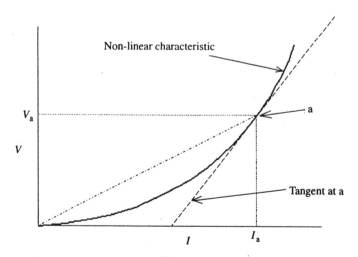

Figure 3.4 A non-linear V/I characteristic

3.2.2 Conductance

The conductance, G, of a resistance is given by

$$G = 1/R \qquad (3.6)$$

The unit of conductance is the siemen (S).

In parallel circuits conductance can be a more convenient parameter to use than resistance. This is demonstrated in later examples.

3.2.3 Capacitance

Again the modelling approach of circuit analysis is emphasized. The physical form of the capacitance is not of prime importance. An ideal capacitor may be regarded as a component that can store energy (in an electric field) but not dissipate energy. A parallel plate capacitor provides a convenient image to consider although this is not a necessary restriction.

In Figure 3.5, the capacitance, C, of a capacitor is defined by the expression

$$Q = VC \qquad (3.7)$$

A capacitance of 1 farad (F) will store a charge of 1 coulomb at a potential difference of 1 volt.

To change the voltage across a capacitance the charge on the capacitance must be changed. This implies a movement of charge into or out of the capacitance; that is, a current flow. Using lower case letters for quantities that are varying, a current i equal to dq/dt must flow and since $q = vC$ we have

$$i = dq/dt = C \, dv/dt \qquad (3.8)$$

where lower case letters denote instantaneous values.

This provides a relationship for current in terms of voltage for the capacitance. Note that this is a positive expression with the conventional labelling of current and voltage as shown in Figure 3.6.

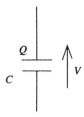

Figure 3.5   A capacitance

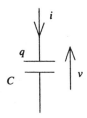

Figure 3.6 Sign convention for voltage and current

Integrating both sides of the equation $i = dq/dt$ gives $q = \int i \, dt$. It is thus also possible to express the voltage across the capacitance in terms of the current:

$$v = (1/C) \int i \, dt \tag{3.9}$$

To increase q quickly a large current must be applied as suddenly as possible. As the integral of a step function of current is a linear ramp waveform, it can be deduced that for any finite value of current the voltage across a capacitor can never change instantaneously. This has important application in the transmission of pulses in electronic circuits.

The instantaneous power in a capacitance is

$$p = vi$$
$$= vC \, dv/dt$$

Thus for a constant applied voltage the power dissipated in a capacitance is zero.

If the voltage changes, energy moves into or out of the electric field. Over any cycle of change that ends with the voltage returning to its original value the mean power dissipated is zero.

The energy stored in a capacitance may be calculated as follows: assume zero energy at $t = 0$ and that at some time, τ, the voltage across the capacitance is V. Since power is the rate of change of energy the total energy stored at time τ is given by

$$\text{energy} = \int_0^\tau p \, dt$$
$$= \int_0^\tau vi \, dt$$
$$= \int_0^\tau vC(dv/dt)dt$$
$$= C \int_0^V v \, dv$$
$$= C[v^2/2]_0^V$$
$$= CV^2/2 \tag{3.10}$$

3.2.4 Inductance

An ideal inductor may be regarded as a component that can store energy (in a magnetic field) but not dissipate energy.

In Figure 3.7 the inductance, L, of an inductor is defined by

$$v = L \, \mathrm{d}i/\mathrm{d}t \qquad\qquad (3.11)$$

An inductance of 1 henry (H) will produce a voltage drop of 1 volt when the current changes at a rate of 1 ampere per second.

Integrating both sides gives

$$i = (1/L) \int v \, \mathrm{d}t \qquad\qquad (3.12)$$

It follows that for any finite value of voltage the current through an inductance cannot change instantaneously.

The thoughtful reader may suggest that if the current-carrying circuit is broken, by throwing a switch or cutting a wire, then the current will change very quickly. Experiment shows that the voltage across the inductance will rise to a very large value, large enough to exceed the breakdown voltage of the surrounding air and cause a spark.

The instantaneous power in an inductance is

$$p = iv$$
$$= iL \, \mathrm{d}i/\mathrm{d}t$$

For any constant current the power dissipated is zero.

When the current changes the energy stored in the magnetic field changes. Over any cycle of change that ends with the current returning to its original value the mean power dissipated is zero.

The energy stored in an inductance may be calculated as follows: assume zero energy at $t = 0$ and that at some time, τ, the current through the inductance is I. Since power is the rate of change of energy the total

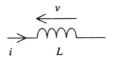

Figure 3.7 An inductance

energy stored at time τ is given by

$$\text{energy} = \int_0^\tau p\,\mathrm{d}t$$

$$= \int_0^\tau iL(\mathrm{d}i/\mathrm{d}t)\mathrm{d}t$$

$$= L\int_0^I i\,\mathrm{d}i$$

$$= L[i^2/2]_0^I$$

$$= LI^2/2 \qquad\qquad (3.13)$$

3.3 Sources

Sources can be categorized as follows:

Independent	or	controlled
Ideal	or	practical
Voltage	or	current

An independent voltage source is a two-terminal device where the voltage is independent of any external factor. A dependent or controlled voltage source has a voltage that is controlled by the voltage or current in another part of the circuit. It is possible to have a voltage-controlled voltage source, a current-controlled voltage source, a voltage-controlled current source, and a current-controlled current source. Controlled sources occur in electronic systems, but in this introduction to circuit analysis only independent sources will be considered. All sources used in circuit diagrams to be analysed will be ideal sources. Practical sources will be modelled using ideal sources and passive components.

3.3.1 The ideal voltage source

The voltage between the terminals in Figure 3.8 is held at V volts whatever the current drawn. If a resistance is connected across an ideal

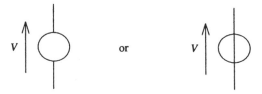

Figure 3.8 The ideal voltage source

voltage source and its value is reduced, the current drawn will increase. In the limit a short circuit will result in an infinite current flowing with a finite voltage V across the short circuit. This is a theoretical situation but should be well understood. Ohm's law still holds as mathematically infinity multiplied by zero is indeterminate, i.e. it can be equal to the finite value V.

3.3.2 The ideal current source

The current through the source in Figure 3.9 will be maintained at I amps whatever the voltage developed across it. In the theoretical limit of a current I flowing through an open circuit the voltage developed across the current source is infinite.

Polarity conventions

It will be observed in Figure 3.10 that, while for the passive components positive current enters at the positive end of the component, for sources the opposite is true.

Figure 3.9   The ideal current source

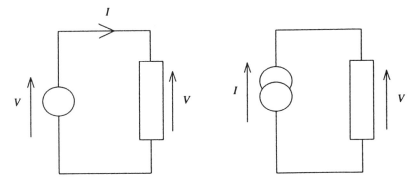

Figure 3.10   Polarity conventions

3.4 Ohm's Laws and Kirchhoff's Laws

Ohm's law and Kirchhoff's laws are the basic laws of circuit analysis. In principle all problems could be solved by directly applying these laws. Additional theorems and formalized analysis techniques are, however, very desirable and are developed in the next chapter.

3.4.1 Ohm's law

This has already been introduced. It may be expressed in words as:

The voltage drop across a resistance is directly proportional to the current through it.

It may be written or used in three forms:

$$V = IR \qquad I = V/R \qquad R = V/I \tag{3.14}$$

It applies to all *linear* resistances.

3.4.2 Kirchhoff's laws

Kirchhoff's voltage law states:

The algebraic sum of the voltages around any closed loop of a circuit is zero.

This is quite general and does not distinguish between voltages across passive components and voltages across sources. Care is required in getting the polarities correct and the conventions described earlier must always be used.

Kirchhoff's current law states:

The algebraic sum of the currents entering any node of a circuit is zero.

(A node is a point in the circuit; see section 4.1.2 for a full definition.)

This may be alternatively expressed as the sum of the currents leaving any node is zero or as the sum of the currents entering any node is equal to the sum of the currents leaving that node.

The current law implies that it is not possible to have a source or sink of electrons at a node in the circuit. It should be noted that this applies to electronic circuits as well as to passive circuits.

3.5 Combining components in series and parallel

3.5.1 Resistances

By direct application of Kirchhoff's laws it may be shown that for resistances connected in series the total resistance R_t is the sum of the individual resistances.

$$R_t = R_1 + R_2 + R_3 + \ldots \tag{3.15}$$

For resistances connected in parallel the total resistance R_t is given by

$$1/R_t = 1/R_1 + 1/R_2 + 1/R_3 + \ldots \tag{3.16}$$

As the most commonly occurring parallel combination is for just two resistances it is convenient to write this as

$$R_t = (R_1 R_2)/(R_1 + R_2) \tag{3.17}$$

That is, the total resistance is given by the *product* of the two individual resistances divided by their *sum*.

3.5.2 Inductances and capacitances

These components become important in a.c. analysis, treated in Chapter 5. However, it is appropriate to deal with their series and parallel combination here. Using the current/voltage relationships previously introduced and applying Kirchhoff's laws it may be shown that inductances obey the same rules as resistances. For inductances in series the total inductance is the sum of the separate inductances. For inductances in parallel the total inductance is the reciprocal of the sum of the separate reciprocals.

For capacitances in parallel the total capacitance is the sum of the separate capacitances, as might be intuitively expected. For capacitances in series the total capacitance is the reciprocal of the sum of the reciprocals of the separate capacitances.

Summarizing:

For capacitances in parallel	$C_t = C_1 + C_2 + C_3 + \ldots$
For capacitances in series	$1/C_t = 1/C_1 + 1/C_2 + 1/C_3 + \ldots$
For inductances in parallel	$1/L_t = 1/L_1 + 1/L_2 + 1/L_3 + \ldots$
For inductances in series	$L_t = L_1 + L_2 + L_3 + \ldots$

3.5.3 Sources

Sources must be treated with care. For ideal voltage sources in series the total voltage is the algebraic sum of the separate voltages.

It is not valid to have two ideal voltage sources of different value connected in parallel as it is impossible to have two different values for the voltage between the same two points. The model of a practical source will include some resistance and hence practical sources may be connected in parallel although large currents may flow between the sources.

It is sometimes sensible to use two practical sources of the same value in parallel to provide a greater current to the rest of the circuit than can be drawn from either source individually.

For ideal current sources in parallel the total current is the algebraic sum of the separate currents.

It is not valid to connect in series two ideal current sources of different value as Kirchhoff's current law would be violated. For practical current sources arguments similar to those used for practical voltage sources may be applied.

The following two examples serve to test your application of the material covered so far. The answers, with explanation, are given after the end of the chapter problems. Make a serious attempt at them before referring to the answers.

Example 3.1

Calculate the current, I, through the resistor in Figure 3.11.

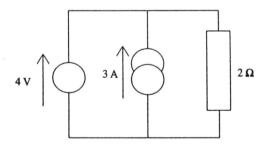

Figure 3.11 Circuit for example 3.1

Example 3.2

Calculate the voltage, V, across the resistor in Figure 3.12.

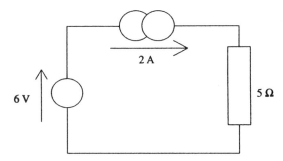

Figure 3.12   Circuit for example 3.2

_____ **Problems** _____

P3.1   A positive charge of 10 coulombs passes in one second through a wire in a direction from a to b.
   (i)   What is the current if the assumed reference direction is from a to b?
   (ii)   What is the current if the assumed reference direction is from b to a?
   (iii) How would these answers change if the moving charges were negative instead of positive?

P3.2   In moving from a to b, a coulomb of charge changes its energy by 10 joules. State the voltage of point a with respect to point b if:
   (i)   the charge is positive and the energy is lost,
   (ii)   the charge is positive and the energy is gained,
   (iii) the charge is negative and the energy is lost,
   (iv)   the charge is negative and the energy is gained.

P3.3   In a certain two-terminal device a positive current of 10 amperes enters at terminal a and leaves from terminal b. What is the power absorbed in the device when:
   (i)   the voltage at a is 10 V positive with respect to b?
   (ii)   the voltage at b is 10 V positive with respect to a?
   (iii) the voltage at a is −10 V with respect to b?

P3.4   Which of the devices shown in Figure P3.1 are sources? The numbers indicate current and voltage values in amperes and volts.

P3.5   Points a, b, c, and d are junctions in an electric network. The following voltages are known:

$$V_{ab} = 10 \quad V_{bc} = -2 \quad V_{ad} = 5$$

Calculate V_{cd}.

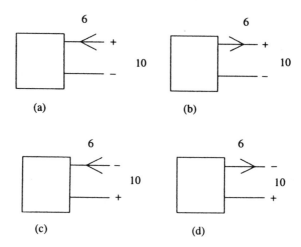

Figure P3.1

P3.6 Points a, b, c, d, and o are junctions in an electric network. The following currents are known:

$$I_{ao} = 4 \quad I_{bo} = -3 \quad I_{co} = 10$$

Calculate I_{do}.

P3.7 A voltage source of 10 volts is connected in series with a resistance of 5 ohms. Calculate the current in the circuit and the power absorbed by the resistance.

P3.8 A voltage source of 6 volts is connected across a capacitor of capacitance 1 microfarad. Calculate the energy stored in the capacitor.

P3.9 A current of 10 amperes flows through an inductor of inductance 1 henry. Calculate the energy stored in the inductor.

P3.10 A current described by the graph shown in Figure P3.2 flows through an inductance of 500 mH. Deduce an expression for the voltage drop across the inductance during the time $t = 0$ to $t = 2$.

P3.11 The current of problem P3.10 now flows into a 100 nanofarad capacitance. Deduce an expression for the voltage developed across the capacitor in the interval $t = 0$ to $t = 2$.

P3.12 For the situation given in problem P3.8, *describe* what happens to the energy when the voltage source is removed.

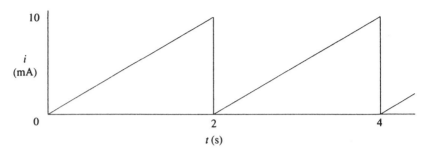

Figure P3.2

P3.13   For the situation given in problem P3.9, *describe* what happens to the energy and the current flow if the current source is replaced by a short circuit.

P3.14   How much power is delivered by an ideal voltage source of 10 volts if:
(i)   it is open-circuited?
(ii)   a resistance of 1 ohm is placed across it?
(iii)   it is short-circuited?
(iv)   it is connected in parallel with a current source of 1 ampere and direction such that the current enters the positive terminal of the voltage source?

P3.15   How much power is delivered by an ideal 10 ampere current source if:
(i)   it is open-circuited?
(ii)   it is short-circuited?
(iii)   a resistance of 1 ohm is placed in series with it?
(iv)   a voltage source of 1 volt is placed across it such that the current enters the negative terminal of the voltage source?

———— *Answers to the two examples set previously* ————————

3.1   The ideal voltage source maintains 4 V across the resistor and therefore the current through the resistor is 2 A. As 3 A is supplied by the current source a current of 1 A must flow into the voltage source. The voltage source here is dissipating power, not supplying power.

3.2   The current source establishes the current through the resistor as 2 A. The voltage across the resistor is therefore 10 V. The voltage across the current source is 4 V. The current source here is providing a power of 8 W, the voltage source providing 12 W, and the resistor dissipating 20 W.

CHAPTER FOUR

Circuit analysis theorems and techniques

This chapter develops the application of Kirchhoff's laws to the analysis of direct current (d.c.) networks and introduces several theorems of importance. Networks of resistances and independent sources are considered. If a good understanding of these techniques and theorems can be achieved then analysis of general alternating current (a.c.) circuits becomes easier.

4.1 Definition of terms

The networks to be analysed consist of ideal components connected together in particular ways. The description of the form of the network connections is known as the network topology. It is necessary to have clear definitions of the terms to be used. For simplicity only those terms used in this book are defined here.

4.1.1 An element

An element of a network is a single ideal component.

In this chapter elements are resistances, voltage sources, or current sources. When a.c. circuits are introduced then elements could also be capacitances or inductances.

4.1.2 A node

A node is the junction point of two or more elements.

The junction point of two elements is usually a trivial node. If the two elements are both passive they can be combined into a single effective element. If one is a source then either the current or voltage is known and does not require a calculation. For this reason some texts refer to a node only when it is the junction of more than two elements. We will use the two or more definition.

4.1.3 A loop

A loop is *any* closed path through the network that encounters each node along the path once only.

4.1.4 A mesh

A mesh is a loop that cannot be divided into smaller loops, i.e. a loop having no branches in its interior.

4.1.5 Illustration of definitions

These definitions are crucial to the analysis techniques to be introduced. Figure 4.1 illustrates the definitions.

Points a, b, c, and d are nodes of the network. The paths listed below are all loops:

$$a \rightarrow b \rightarrow d \rightarrow a$$
$$a \rightarrow b \rightarrow c \rightarrow d \rightarrow a$$
$$b \rightarrow c \rightarrow d \rightarrow b$$
$$a \rightarrow c \rightarrow b \rightarrow a$$
$$a \rightarrow c \rightarrow d \rightarrow a$$

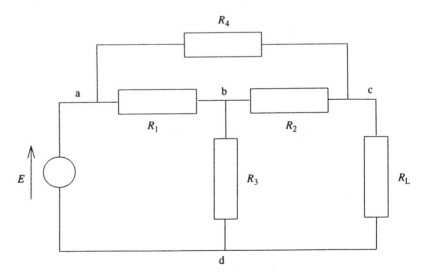

Figure 4.1   Defining meshes, loops, and nodes

The closed path a → b → d → c → b → d → a is not a loop as it encounters nodes b and d twice.

The only meshes in Figure 4.1 are

$$a \rightarrow b \rightarrow d \rightarrow a$$
$$a \rightarrow c \rightarrow b \rightarrow a$$
$$b \rightarrow c \rightarrow d \rightarrow b$$

4.2 Voltage and current dividers

4.2.1 The voltage divider

A commonly occurring circuit is shown in Figure 4.2.

Provided that no external current flows from the point A, the current in R_1 must flow through R_2. Thus applying Ohm's law to R_2 and to R_1 and R_2 in series:

$$V_b = IR_2$$
$$V_a = IR_1 + IR_2 = I(R_1 + R_2)$$

hence

$$V_b/V_a = R_2/(R_1 + R_2) \tag{4.1}$$

This result can be expressed in words as 'the ratio of the voltage across R_2 to the voltage applied across R_1 and R_2 in series is equal to the ratio of the resistance of R_1 to the resistance of R_1 and R_2 in series'.

By suitable choice of R_1 and R_2, V_b can be made equal to any fraction of V_a. The arrangement is therefore called a *voltage divider*.

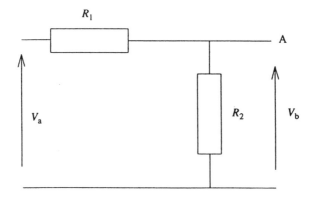

Figure 4.2 The voltage divider

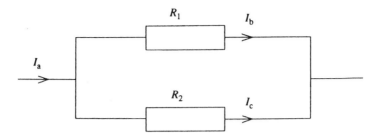

Figure 4.3   The current divider

4.2.2 The current divider

An equally useful relationship can be deduced for the currents in a parallel circuit as shown in Figure 4.3.

As the voltage is the same across each resistance, applying Ohm's law gives

$$I_b R_1 = I_c R_2 \quad \text{or} \quad I_b/I_c = R_2/R_1$$

Adding 1 to each side of the equation leads to

$$(I_b + I_c)/I_c = (R_1 + R_2)/R_1$$

But $I_b + I_c = I_a$, giving

$$I_c/I_a = R_1/(R_1 + R_2) \tag{4.2}$$

This result is also best remembered by expressing it in words. The ratio of the current in either branch of a parallel circuit to the total current flowing into the circuit is equal to the ratio of the resistance of the *other* branch to the sum of the resistances in both branches.

4.3 Mesh and nodal analyses

All linear circuits can be analysed by applying Ohm's law and Kirchhoff's laws; however, there are advantages to be gained by applying these laws in a structured manner. Consider by way of example the circuit of Figure 4.1. If values for the resistances and for the source voltage are known then all the circuit currents and voltages may be calculated. There are five unknown voltages and six unknown currents. However, by writing a Kirchhoff current law (KCL) equation for each node and a Kirchhoff voltage law (KVL) equation for every loop 16 valid equations are possible; that is, five more than required. It is desirable to produce a necessary and sufficient number of equations. Mesh and nodal analyses are formalized ways of applying the laws to achieve this.

4.3.1 Mesh analysis

This technique concentrates on calculating the currents in the circuit. Voltages may then be obtained by independent application of Ohm's law. To produce a necessary and sufficient number of equations *only* meshes (not loops) are considered.

The concept of a mesh current is best shown by an example. Consider the circuit of Figure 4.1 which is redrawn in Figure 4.4 with the mesh currents labelled.

The current I_1 is considered to circulate around the mesh containing R_1, R_4, and the source. Call this mesh 1 in order to refer to it in the text; similarly the mesh associated with I_2 is called mesh 2, and so on. The actual current flowing in any element is the algebraic sum of the mesh currents flowing in that element. For example:

the current in R_4 is $(I_1 - I_3)$ downwards
the current in R_L is simply I_3

KVL equations should now be written for all meshes for which the mesh current is unknown (as will be seen later a mesh current may have a known value if the mesh contains a current source):

For mesh 1 $\qquad R_1(I_1 - I_2) + R_4(I_1 - I_3) = E$

For mesh 2 $\quad R_3 I_2 + R_2(I_2 - I_3) + R_1(I_2 - I_1) = 0$

For mesh 3 $\quad R_2(I_3 - I_2) + R_L I_3 + R_4(I_3 - I_1) = 0$

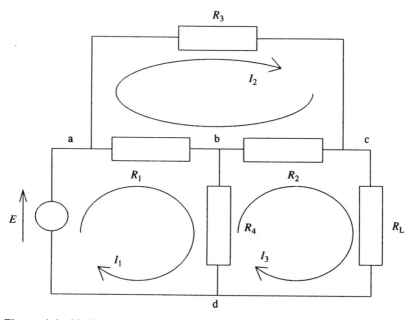

Figure 4.4 Mesh currents

These equations are a direct statement of the KVL; each is the sum of the individual voltage drops across each element in the mesh. This is the essence of mesh analysis. The reader may now go to the problems at the end of the chapter and practise writing the KVL equations. The next section describes a convenient way to write the equations with the unknown terms already collected together. This will be found useful and time saving but it does not add any new electrical concept or understanding.

4.3.2 Gathering the terms

To solve the equations the terms must be gathered together. Doing this for the mesh 1 equation gives

$$(R_1 + R_4)I_1 - R_1 I_2 - R_4 I_3 = E$$

A form for this equation may be observed. The first term is the mesh current multiplied by the sum of the values of all the resistances in the mesh. Now for every resistance containing a further mesh current there is a subtracted term equal to the resistance value multiplied by the further mesh current. The total is equated to the sum of the voltage sources in the mesh.

This provides a convenient way of writing the equations with the unknown terms already gathered for solution. It is completely general provided the mesh currents are *all labelled as flowing in the same direction*.

The sum of the values of all the resistances in a mesh is called the mesh resistance. A resistance that is common to two meshes is said to couple the meshes. The resistance is called the coupling resistance and one of the currents is called the coupled mesh current.

_____ **Example 4.1** _____

Calculate the current in the 3 ohm resistance of Figure 4.5.

Label the mesh currents. The current in the right hand mesh is determined by the current source and is labelled accordingly.

For mesh 1 $12I_1 - 4I_2 = 14$ (i)

For mesh 2 $-4I_1 + 12I_2 - 5(-2) = 0$ (ii)

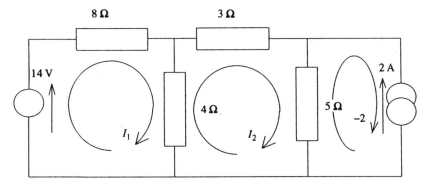

Figure 4.5 Circuit for example 4.1

Adding (i) + 3 times (ii) gives

$$32I_2 = -16$$
$$I_2 = -16/32 = -0.5 \text{ A}$$

The significance of the minus sign is that the half ampere current in the 3 ohm resistance flows in the opposite direction to that indicated by the I_2 label; that is, it flows from right to left.

4.3.3 Nodal analysis

This technique calculates the voltages at the circuit nodes. The currents are subsequently obtained using Ohm's law.

Choose any node of the circuit as the reference node and label it as such. Label the voltage, with respect to the reference node, at each other node. Write a KCL equation for each node at which the voltage is unknown.

_____ **Example 4.2** _____

The circuit in Figure 4.6 is the same circuit as in Figure 4.5. Again calculate the current in the 3 ohm resistance.

At node a $\quad (V_a - 14)/8 + V_a/4 + (V_a - V_b)/3 = 0$ $\qquad$ (i)

At node b $\quad (V_b - V_a)/3 + V_b/5 = 2$ $\qquad$ (ii)

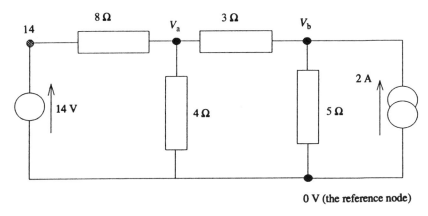

Figure 4.6   Circuit for example 4.2

Multiply (i) by 24 and (ii) by 15 and collect terms together:

$$17V_a - 8V_b = 42$$
$$-5V_a + 8V_b = 30$$

$$12V_a \qquad = 72$$

Whence $V_a = 6$ V giving $V_b = 7.5$ V.
The current in the 3 ohm resistance is $(V_a - V_b)/3 = -1.5/3 = -0.5$ A.

Experience will help the reader to identify which technique will produce the easier solution for a given problem. Both techniques should be mastered. For the circuits commonly met nodal analysis will probably be the better solution more often than mesh analysis.

4.3.4 More on nodal analysis

This section may be omitted at a first reading; however, it will be found to be of value when the reader is confident of his or her ability to apply nodal analysis as introduced in the previous section.

Experience will show that a node that is the junction of only two elements is a trivial node in this context and may generally be left unlabelled to reduce the number of unknowns. An exception is when one of the elements is an ideal voltage source. No harm will come from labelling any node, so if there is any uncertainty then label it.

It might be expected that the nodal equations could be written with the unknown terms already gathered as was the case for mesh analysis. The form for the equations is described below but it is left as an exercise for the reader to show that this is generally valid.

For each node at which the voltage is unknown write an equation as follows:

Start with a term

(node voltage) × (sum of the conductances connected to the node)

For every node linked to the node in question include a subtracted term: (linked node voltage)

× (the conductance of the element linking the nodes)

On the other side of the equation sum all current sources flowing directly into the node. For the circuit of example 4.2 the equations formed in this way are

$$V_a(1/8 + 1/4 + 1/3) - V_b(1/3) - 14(1/8) = 0$$

$$-V_a(1/3) + V_b(1/3 + 1/5) = 2$$

4.4 Thévenin's theorem and Norton's theorem

These theorems are quite powerful and although they are deliberately limited in the following sections to passive components and independent sources, they are used extensively in electronics. Their application is most appropriate in situations where an initial calculation can result in a simple circuit which is equivalent to the original more complex circuit. Subsequent repetitive calculations, such as to ascertain the effects of varying frequency or varying load resistance, are then greatly simplified. For a single analysis of a given circuit the calculation of the Thévenin or Norton circuit can be as involved as the complete solution to the problem. However, such calculations do provide the student with practice in the use of the theorems and allow the results to be confirmed by analysis techniques not involving these theorems.

4.4.1 Thévenin's theorem

Consider a network composed of linear passive elements and independent sources. To investigate the conditions in any passive part of the network, Thévenin's theorem states that the rest of the network may be replaced by an equivalent circuit consisting of an ideal voltage source in series with a resistance.

In Figure 4.7 network A is any network of linear passive components and independent sources. Thévenin's theorem states that the circuit within

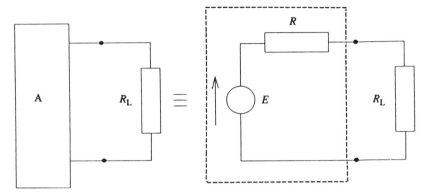

Figure 4.7   Illustrating Thévenin's theorem

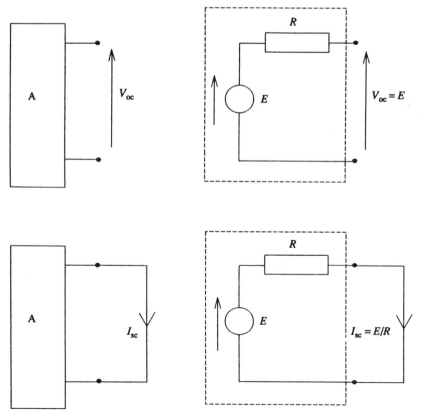

Figure 4.8   Calculating Thévenin values

the dashed lines is equivalent to network A in so far as the current in, or voltage across, R_L is concerned.

Values must be calculated for E and R, the Thévenin voltage and Thévenin resistance respectively.

Choosing any arbitrary value for R_L it must be possible to calculate the current and voltage for R_L when it is connected to network A. For the Thévenin circuit to be equivalent the same values must exist when R_L is connected to the Thévenin circuit. Thus an equation is obtained in E and R_L. Repeating this for any other value of R_L gives a second equation which is necessarily independent. Therefore the two equations may be solved for E and R_L.

This calculation is simplified if the values chosen for R_L are the limit values of zero and infinity, corresponding to short-circuit and open-circuit terminations for the network.

The steps to calculate E and R_L are illustrated in Figure 4.8 and may be stated as

1. Calculate the open-circuit terminal voltage for the network. This must be the required value of E as there is no voltage drop across R when the Thévenin equivalent circuit is open circuit.
2. Short-circuit the output of the network and calculate the current in the short circuit. This current must be equal to E/R and hence R may be calculated.

—— **Example 4.3** ————————————————————————

Calculate a Thévenin equivalent for the circuit of Figure 4.9 to the left of terminals A and B.

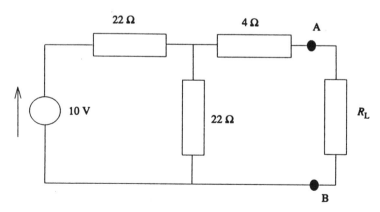

Figure 4.9 Circuit for example 4.3

Remove R_L and calculate the voltage between A and B. There is no current through the 4 ohm resistance when the network is open circuit; hence the voltage at A is the same as the voltage at the junction of the two 22 ohm resistances. V_{AB} is thus equal to 5 V.

Replace R_L by a short circuit as shown in Figure 4.10 and calculate the current through the short circuit, i.e. the current through the 4 ohm resistance.

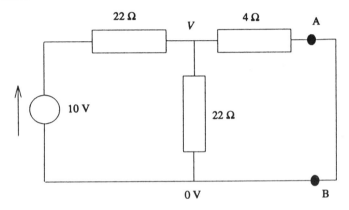

Figure 4.10   Calculating the short-circuit current

Using nodal analysis

$$V(1/22 + 1/22 + 1/4) - 10(1/22) = 0$$

giving $V = 4/3$ V and hence the current in the 4 ohm resistance is $1/3$ A. The Thévenin equivalent resistance is thus 5 divided by $1/3$ which is 15 ohms. The Thévenin equivalent circuit is therefore as shown in Figure 4.11.

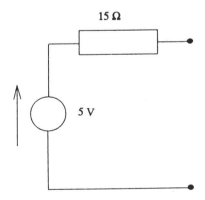

Figure 4.11   The equivalent circuit for example 4.3

4.4.2 Internal resistance of ideal sources

There is an alternative method of calculating R for the Thévenin circuit. To use this it is important to realize that the internal resistance of an ideal voltage source is zero and that of an ideal current source is infinite. The following argument may help confirm the above statement.

Consider a practical voltage source of internal resistance r and open-circuit voltage E. If a current is drawn from the source the terminal voltage will fall owing to the voltage drop across r. The smaller the value of r the less the change in terminal voltage. In the limit of $r = 0$ the terminal voltage will remain equal to E whatever the current drawn, which is the definition of an ideal voltage source. A similar argument can be used for the ideal current source.

To calculate the Thévenin equivalent resistance for a circuit all the sources in the network should be replaced by their internal resistances. The total effective resistance between the network terminals is then calculated. This is the value required for the Thévenin resistance. In practice this method is usually much easier than calculating the short-circuit current.

_____ **Example 4.4** _____

Use this method for the circuit of example 4.3.

Replacing the ideal voltage source by its internal resistance, i.e. a short circuit, the resistance between the terminals A and B is that of the 4 ohm resistance in series with the parallel combination of the two 22 ohm resistances. That is, the Thévenin equivalent resistance, R, is given by

$$R = 4 + 11 = 15 \text{ ohms}$$

4.4.3 Norton's theorem

Consider a network composed of linear passive elements and independent sources. To investigate the conditions in any passive part of the network, Norton's theorem states that the rest of the network may be replaced by an equivalent circuit consisting of an ideal current source in parallel with a conductance.

In Figure 4.12 values for I and G may be obtained by measuring or calculating the open-circuit voltage and the short-circuit current of the network to be replaced.

The short-circuit current is seen to be the value required for the Norton ideal current source and the open-circuit voltage of the Norton circuit is clearly I/G. Thus, knowing I, G can be calculated.

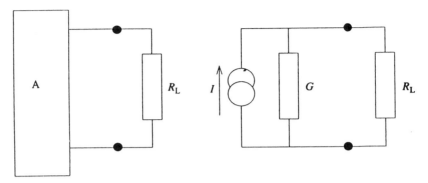

Figure 4.12   Illustrating Norton's theorem

In practice it is usually more convenient to calculate G by replacing all the sources in the network by their internal resistances and calculating the conductance between the terminals.

Although it is customary to define the Norton circuit in terms of a parallel conductance it is of course quite acceptable to refer to a parallel resistance, R, where $R = 1/G$.

_____ **Example 4.5** _____

Calculate a Norton equivalent for the circuit of Figure 4.9 to the left of terminals A and B.

The first step is to calculate the equivalent resistance (or conductance). This is exactly the same calculation as in example 4.4 giving a Norton resistance of 15 ohms (or a conductance of 1/15 siemens).

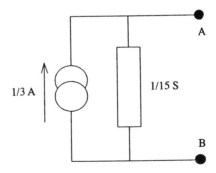

Figure 4.13   Norton equivalent for example 4.5

To calculate the value of the Norton ideal current source the terminals A and B of the original circuit must be shorted together and the current in this short circuit calculated. This has already been done; the circuit is shown in Figure 4.10 and the short-circuit current was calculated as 1/3 A. Thus the Norton equivalent circuit for the circuit of Figure 4.9 is as shown in Figure 4.13.

4.4.4 Conversion of sources

If a given network may be replaced by a Thévenin equivalent circuit and the same network may also be replaced by a Norton equivalent circuit then it follows that the Thévenin and Norton circuits are equivalent, one to the other.

It must thus be possible to replace any practical voltage source by an equivalent current source and vice versa. This is illustrated in Figure 4.14.

To convert to a voltage source:

$$E = I/G \quad \text{and} \quad R = 1/G$$

To convert to a current source:

$$I = E/R \quad \text{and} \quad G = 1/R$$

This can sometimes be a useful step in simplifying the analysis of a circuit. Note that it is not possible to replace an ideal voltage source by an ideal current source, or vice versa.

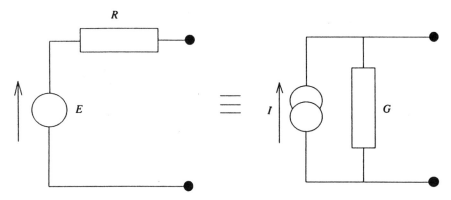

Figure 4.14 Conversion of sources

4.5 Superposition theorem

In a linear circuit if a stimulus S_a *produces a response* R_a *at a given place in the circuit and a stimulus* S_b *produces a response* R_b *at the same place, then the response at that place to both stimuli applied together is* $R_a + R_b$.

This follows directly from the concept of linearity and may be demonstrated by an example.

Consider the problem posed in Figure 4.5. To apply the superposition theorem to this problem the current in the 3 ohm resistance due to the current source alone is calculated. The current in the 3 ohm resistance due to the voltage source alone must then be calculated. Finally these currents are combined algebraically to find the total current in the 3 ohm resistance.

To calculate the current due to the current source, the voltage source must be removed as a stimulus but the circuit path it provided must be preserved. This is achieved by replacing the voltage source by a path containing its internal resistance. As for an ideal voltage source the internal resistance is zero this is always a simple step: that is replace the voltage source by a short circuit. The circuit therefore becomes as shown in Figure 4.15.

The easiest way to calculate I_a is to calculate the resistance of the path consisting of the 3 ohm resistance in series with the parallel combination of the 4 ohm and 8 ohm resistances. This is 17/3 ohms. The current from the source splits between this path and the 5 ohm path. Applying the current divider rule I_a is given by

$$I_a = \frac{2 \times 5}{(5 + 17/3)} = 15/16 \text{ A}$$

Now replace the current source by its internal resistance, an open circuit in this case, and calculate the current in the 3 ohm resistance due to the voltage source. The circuit now becomes as shown in Figure 4.16.

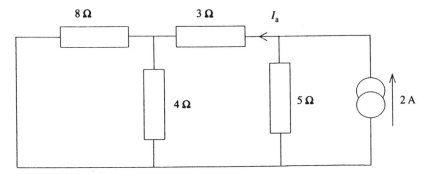

Figure 4.15 Superposition theorem; response due to the current source

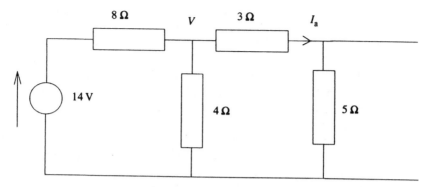

Figure 4.16 Superposition theorem; response due to the voltage source

Applying nodal analysis

$$(V - 14)/8 + V/4 + V/(3 + 5) = 0$$
$$4V/8 = 14/8 \quad \text{or} \quad V = 7/2$$

Hence $I_b = 3.5/(3 + 5) = 7/16$ A.

Since I_a and I_b flow in opposite directions the total current in the 3 ohm resistance is $15/16 - 7/16 = 0.5$ A flowing towards the voltage source.

As would be expected this is the same answer as obtained previously.

The superposition theorem is most useful when the removal of one or more of the sources simplifies the calculation significantly. It is then easier to do several small calculations rather than one large one. It also has application when a complex source can be replaced by the addition of two simpler sources. Practice and experience are needed to judge the best way to tackle a problem. Do not think that any circuit with multiple sources must be analysed using the superposition theorem.

It is appropriate here to offer a word of warning regarding a common mistake which arises when the superposition theorem is used to calculate the power in a component. Because power is proportional to I^2 or to V^2 it is not a linear relationship to the source voltage or current. It is vital therefore to calculate the total current or voltage in the component before calculating the power. Calculating the power due to one source and then the power due to a second source and adding the powers will *not* give a correct answer.

4.6 Tee to Pi conversion

So far whenever it has been required to calculate the total resistance between two points it has been possible to do this by combining resistances in series and in parallel. Consider now the circuit shown in Figure 4.1. The current drawn from the source E could be calculated, in principle, by calculating the total effective resistance between the nodes 'a' and 'd'. However, it can be seen that no two resistances are connected either in series or in parallel. This situation arises in a number of commonly used circuits. It is always possible to calculate the current using mesh or nodal analysis. If required the total resistance can then be found using Ohm's law.

An alternative solution is offered by the possibility of calculating an equivalence between a Tee network and a Pi network. The power engineer will refer to this conversion as a *star* to *delta* conversion. In American texts it is known as *wye* to *delta*.

For both networks of Figure 4.17 to be equivalent the resistances between equivalent pairs of terminals must be equal.

For the Tee:

$$R_{xy} = R_1 + R_2$$
$$R_{xz} = R_1 + R_3$$
$$R_{yz} = R_2 + R_3$$

For the Pi:

$$R_{xy} = R_b(R_a + R_c)/(R_b + R_a + R_c)$$
$$R_{xz} = R_a(R_b + R_c)/(R_b + R_a + R_c)$$
$$R_{yz} = R_c(R_a + R_b)/(R_b + R_a + R_c)$$

For each set of equations add the first two and subtract the third. Thus

$$2R_1 = 2R_aR_b/(R_b + R_a + R_c)$$

or

$$R_1 = R_aR_b/(R_b + R_a + R_c) \tag{4.3}$$

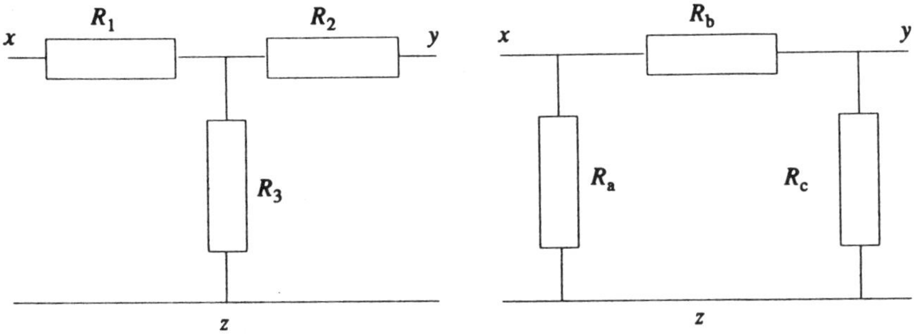

Figure 4.17   Tee to Pi conversion

Similarly

$$R_2 = R_b R_c / (R_b + R_a + R_c) \tag{4.4}$$

and

$$R_3 = R_c R_a / (R_b + R_a + R_c) \tag{4.5}$$

Expressions for the resistance values for the Pi network such that it is equivalent to the Tee may be obtained by manipulating the above expressions or, and this is left as an exercise for the reader, by equating the conductances between corresponding pairs of terminals for each network. The expressions are

$$R_a = (R_1 R_2 + R_2 R_3 + R_3 R_1)/R_2 \tag{4.6}$$
$$R_b = (R_1 R_2 + R_2 R_3 + R_3 R_1)/R_3 \tag{4.7}$$
$$R_c = (R_1 R_2 + R_2 R_3 + R_3 R_1)/R_1 \tag{4.8}$$

_____ **Example 4.6** _____

Calculate the resistance between terminals 'a' and 'b' of the network shown in Figure 4.18.

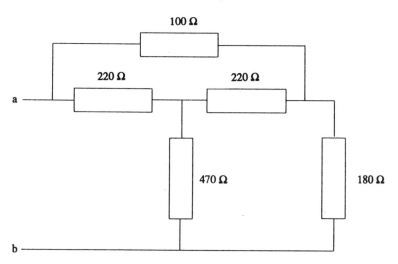

Figure 4.18 Circuit for example 4.6

Replace the Tee of resistors by an equivalent Pi (Figure 4.19).

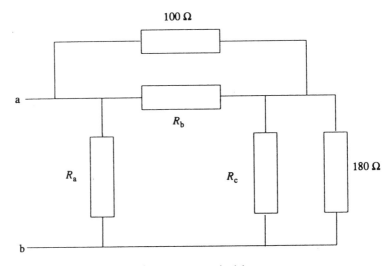

Figure 4.19   First step in solution to example 4.6

Calculate values for R_a, R_b and R_c using equations (4.6)–(4.8).

$$R_a = \frac{220.910}{220} = 1160$$

$$R_b = \frac{220.910}{470} = 542.98$$

$$R_c = \frac{220.910}{220} = 1160$$

R_b is now in parallel with the 100 ohm resistance giving 84.45 Ω.
R_c is in parallel with the 180 ohm resistance giving 155.82 Ω.
The circuit has thus been reduced to a single Pi network and the required resistance between 'a' and 'b' is 199 Ω.

___ **Problems** _____

P4.1   Two batteries of e.m.f. 60 V and 120 V respectively are joined in series to give a 180 V supply across which two resistors of 30 Ω and 240 Ω are joined in series, the 30 Ω resistor being joined at one end to the positive terminal of the 60 V battery. An 80 Ω resistor is connected between the junction of the two resistors and the junction of the two batteries. Calculate the current flowing in each resistor. (Assume that the internal resistance of the batteries is negligible.)

P4.2　A battery has an open-circuit voltage of 250 V and an internal resistance of 1 Ω. Two circuits are connected in parallel to its terminals. Circuit A consists of a cable of resistance 1 Ω, a resistor, and an ammeter connected in series. Circuit B consists of a cable of resistance 2 Ω, a resistor, and an ammeter connected in series. Both ammeters read 5 A. Calculate the voltage at the battery terminals and the voltage across each resistor, assuming that the ammeter resistance can be ignored.

　　　If circuit A is open-circuited, find the new reading of the ammeter remaining in circuit.

P4.3　A 120 V battery is connected across two resistors in series so that a current of unknown value flows. A voltmeter of resistance 200 kΩ is used to measure the voltage across each resistance. The readings are respectively 60 V and 48 V. Calculate the true voltage across the smaller of the two resistors. Assume that the internal resistance of the battery is negligible.

P4.4　Two resistors, A and B, are connected in series across a constant potential d.c. supply. The voltage drops measured respectively across A, B, and the two in series are 20, 50, and 80 volts, the measurement being made by means of an accurate voltmeter of 1000 Ω internal resistance. What are the resistances of A and B?

P4.5　Without converting any of the sources calculate the currents listed below
(a) *by mesh analysis*, (b) *by nodal analysis*:
(i)　the current in the 9 Ω resistance of Figure P4.1,
(ii)　the current in the 9 Ω resistance of Figure P4.2,
(iii)　the current in the 4 Ω resistance of Figure P4.3,
(iv)　the current in the 4 Ω resistance of Figure P4.4.

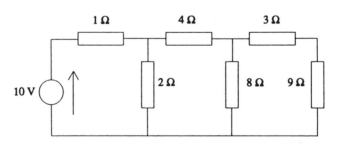

Figure P4.I

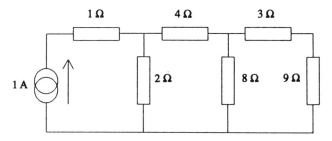

Figure P4.2

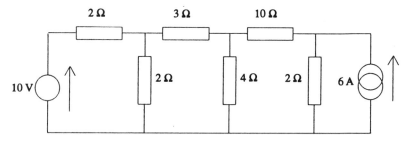

Figure P4.3

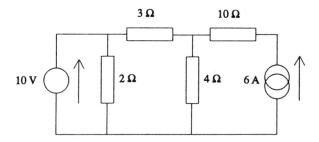

Figure P4.4

P4.6   Calculate, *using the superposition theorem*, the following:
  (i)   the current in the 4 Ω resistance of Figure P4.3,
  (ii)   the current in the 4 Ω resistance of Figure P4.4,
  (iii) the current in the 2 Ω resistance of Figure P4.5,
  (iv)   the current in the 1 Ω resistance of Figure P4.6,
  (v)   the current in the 3 Ω resistance of Figure P4.7.

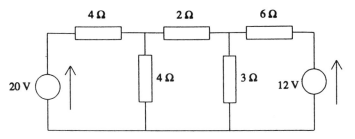

Figure P4.5

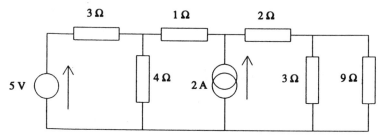

Figure P4.6

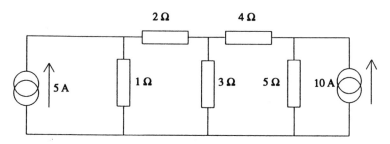

Figure P4.7

P4.7 Convert the voltage sources in Figure P4.8 to equivalent current sources.

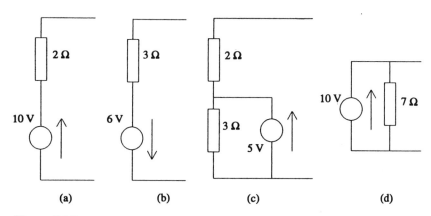

(a)   (b)   (c)   (d)

Figure P4.8

P4.8 Convert the current sources in Figure P4.9 to equivalent voltage sources.

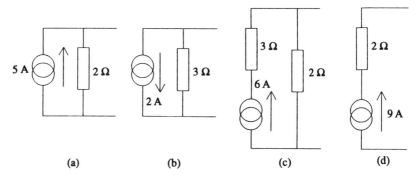

Figure P4.9

P4.9   Convert each of the circuits shown in Figure P4.10 to a single voltage source in series with a single resistance.

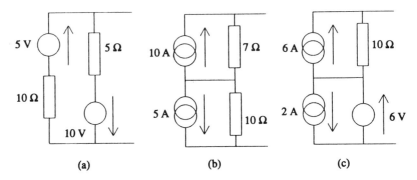

Figure P4.10

P4.10   Convert each of the circuits in Figure P4.11 to a single current source in parallel with a resistance.

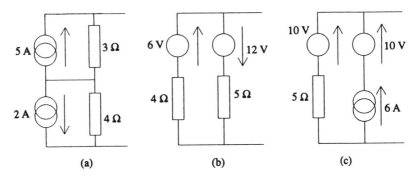

Figure P4.11

In the preceding problems the application of mesh and nodal techniques to differing circuits has been practised and something will have been learned of their relative merits. The superposition theorem has also been applied to the same circuits and the solutions can be compared.

Problems P4.11 to P4.13 apply source conversions to these same circuits and the solutions should be critically compared.

P4.11 For the circuit of Figure P4.1 convert the voltage source to an equivalent current source and calculate the current in the 9 Ω resistance (a) *by nodal analysis*, (b) *by mesh analysis*.

P4.12 For the circuit of Figure P4.5 convert the voltage sources to equivalent current sources and *use nodal analysis* to calculate the current in the 2 Ω resistance.

P4.13 For the circuit shown in Figure P4.3 determine the current in the 4 Ω resistance:

(a) *by mesh current analysis*, having first replaced the current source by an equivalent voltage source,
(b) *by nodal voltage analysis*, having first replaced the voltage source by an equivalent current source.

P4.14 Calculate the current in the 9 Ω resistance of Figure P4.6 by first determining a Thévenin equivalent for the rest of the circuit.

P4.15 Calculate the current in the 3 Ω resistance of Figure P4.7 by first determining a Norton equivalent for the rest of the circuit.

P4.16 For the circuit shown in Figure P4.12 determine the magnitude and direction of the current in the branch AC: (a) *using Thévenin's theorem*, (b) *without using Thévenin's theorem*.

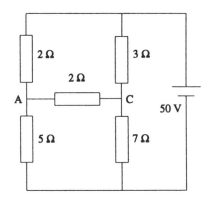

Figure P4.12

P4.17 For the circuit of Figure P4.13 determine the equivalent resistance between terminals A and B.

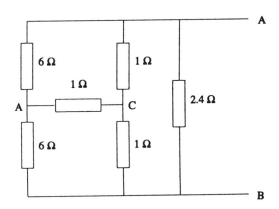

Figure P4.13

Solve the following two problems using any method not involving Thévenin's theorem or Norton's theorem.

P4.18 For the circuit of Figure P4.6 calculate the current in the 9 Ω resistance.

P4.19 For the circuit of Figure P4.7 determine the voltage across the 3 Ω resistance.

Alternating waveforms

This chapter deals with the steady state analysis of a.c. circuits, i.e. circuits in which the current and voltage waveforms are periodic and sufficient time has elapsed from switching on the excitation for any initial transient response to be negligible. This is the most commonly encountered analysis. It is also mathematically simpler than a complete analysis which includes a transient response. For these reasons it is introduced before the more general analysis in Chapter 6. With practice the reader will be able to decide when a steady state solution is adequate and when a complete solution is required. A further reason for considering a steady state analysis separately is that such an analysis becomes particularly straightforward if the excitation is sinusoidal. The importance of the sinusoid is explored and appropriate analysis techniques are dealt with at length. However, in keeping with the book's intention to be fundamental, non-sinusoidal calculations are also covered.

The maximum power transfer theorem is introduced in this chapter as it is the only major theorem that cannot be covered fully by the d.c. only case.

5.1 Definitions

An alternating waveform is a time-varying waveform which periodically reverses its polarity. If after a certain period of time the waveform repeats itself exactly then it is referred to as a *periodic waveform*. The abbreviation a.c. is used to refer to an alternating waveform. The waveform could be precisely described by a mathematical function of time. It would, however, be useful if some properties of the waveform could be described more simply. Consider the waveform shown in Figure 5.1.

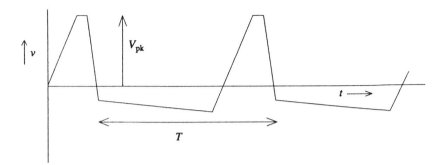

Figure 5.1   A periodic waveform

5.1.1 Peak value

The peak value of a voltage waveform is its greatest deviation from zero. It is normally denoted V_{pk}. This is the value that is easily determined using an oscilloscope. Its practical significance is in relation to the maximum voltage that may be applied to a component. This may be thousands of volts for the breakdown of insulation in a power system or 5 volts for the reverse base–emitter voltage of some transistors.

5.1.2 Average or mean value

An a.c. waveform with no d.c. component has a mathematical average of zero. However, it would be useful to have some general measure of the size of the waveform. This can be achieved by taking the mean of the modulus of the waveform. This value is commonly referred to as simply the mean value – a loose usage of terms but not one to cause the student any lasting difficulty. This mean value is calculated by finding the area under the positive-going part of the waveform and dividing by the baseline for this positive-going part.

5.1.3 Root mean square or effective value

One measure of an a.c. voltage is a comparison with the value of a d.c. voltage that produces the same power in a given load. The a.c. voltage is effectively the same as the d.c. voltage in so far as the power dissipated is concerned.

The instantaneous value of power is v^2/R. As R is a constant the mean power is related to v^2. Hence it is appropriate to calculate the mean value of v^2 and then take the square root of this to make the measure dimensionally correct. The effective value of an a.c. waveform is thus more generally referred to as the *root mean square* or *r.m.s.* value.

The r.m.s. value is the one normally understood when using a.c. quantities. A 5 volt a.c. source implies 5 volts r.m.s.

5.1.4 Period

The period of a waveform, denoted T, is the time taken to go through one complete pattern or cycle of the waveform.

5.1.5 Frequency

The frequency, denoted f, of a waveform is the number of complete cycles occurring in 1 second. The frequency and period are related by

$$f = 1/T$$

5.2 Calculating periodic waveform values

The period, frequency, and peak values of any periodic waveform can normally be obtained by inspection and should not cause any difficulty. Calculation of the mean and r.m.s. values, however, needs a little more thought. The method of calculation is demonstrated in example 5.1.

_____ **Example 5.1** _____

Consider the waveform shown in Figure 5.2. The period is clearly 30 milliseconds.

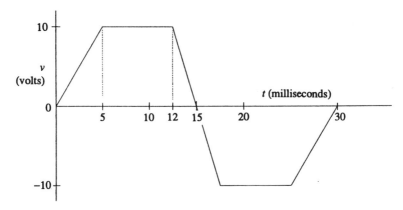

Figure 5.2 Calculation of mean and r.m.s. values

The mean value is taken as the mean value of the positive-going part of the waveform. The mean is the area under the curve divided by the appropriate baseline. In this example the area is calculated quite simply from the geometry of the waveform, but in general the area must be calculated by integration.

For the time interval 0 to 5 ms the area is $10 \times 5 \times 10^{-3} \times 0.5$ units.
For the time interval 5 to 12 ms the area is $10 \times 7 \times 10^{-3}$ units.
For the time interval 12 to 15 ms the area is $10 \times 3 \times 10^{-3} \times 0.5$ units.

The total area is thus 110×10^{-3} units. The appropriate baseline is 0 to 15 ms giving a mean value of $110/15 = 7.33$ V.

The r.m.s. value is the root of the mean square. The first step is therefore to calculate an expression which is the square of the waveform itself. For any waveform with discontinuities in its slope it is necessary to treat the waveform in sections.

For the time interval 0 to 5 ms the waveform is clearly described by the equation

$$v = 2000t$$

Thus the square is

$$v^2 = 4 \times 10^6 t^2$$

This is a parabola and the area must be found by integration. It is good advice to make a freehand sketch of the v^2 waveform as it reduces the chance of making mistakes in calculating areas. In this case the area under the curve is given by

$$4 \times 10^6 \int_0^{5.10^{-3}} t^2 \, dt = 4 \times 10^6 [t^3/3]_0^{5.10^{-3}} = (500/3) \times 10^{-3}$$

For the time interval 5 to 12 ms v^2 is constant at 100 and the area under the curve is simply $100 \times 7 \times 10^{-3}$.

For the interval 12 to 15 ms v is part of the straight line whose equation is

$$v = -(10\,000/3)t + 50$$
Thus $v^2 = (10^8/9)t^2 - (10^6/3)t + 2500$

The area under the curve is therefore

$$\int_{12.10^{-3}}^{15.10^{-3}} [(10^8/9)t^2 - (10^6/3)t + 2500] dt$$

$$= [(10^8/9)t^3/3 - (10^6/3)t^2/2 + 2500t]_{12.10^{-3}}^{15.10^{-3}}$$

$$= 0.1$$

The total area under the v^2 curve is thus

$$0.1666 + 0.7 + 0.1 = 0.9666$$

The mean square is therefore

$$0.9666/15 \times 10^{-3} = 64.44$$

Taking the square root gives the r.m.s. value as 8.028 V.

The shape of the original waveform will significantly affect the mathematics of these calculations but the steps are exactly the same whatever the waveform. The problems at the end of the chapter offer some practice with different practical waveforms and the following section provides a further example where the r.m.s. value of the sinusoid is calculated.

5.3 The sinusoid

Much of the a.c. analysis in this section of the book, as in most other textbooks, is devoted to sinusoids. It is appropriate to consider why this is so. First, as most electrical power is generated by rotating electromagnetic machines the sinusoid is the easiest waveform to generate. Secondly the sinusoid is the only periodic waveform for which the integral and the derivative are of the same form as the waveform itself. This is important if it is considered that a fundamental analysis of most circuits results in a integro-differential equation. Finally the work of Fourier suggests that any periodic waveform may be represented by the sum of a series of harmonically related sinusoids.

It is also appropriate at this point to warn the reader to remember that many of the results derived for sinusoidal waveforms do not apply directly to non-sinusoidal waveforms.

Consider the sinusoid $v = V \sin \theta$ or $v = V \sin \omega t$ shown in Figure 5.3.

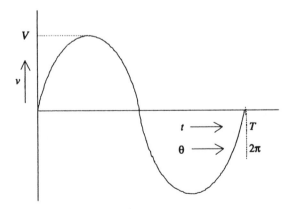

Figure 5.3 The sinusoid

The frequency f is expressed in hertz (Hz). The angular frequency, ω, is expressed in radians per second. As the sinusoid passes through 2π radians in one complete period then clearly

$$\omega = 2\pi f$$

To represent a sine wave shifted in time some initial phase angle must be included:

$$v = V \sin(\omega t + \phi)$$

If $\phi = \pi/2$ radians then $v = V \cos \omega t$.

5.3.1 The mean value of a sinusoid

$$v = V \sin \theta$$

$$V_{mean} = (1/\pi) \int_0^\pi V \sin \theta \, d\theta$$

$$= (V/\pi)[-\cos \theta]_0^\pi$$
$$= (V/\pi)[1 + 1]$$
$$= 2V/\pi$$
$$V_{mean} = (2/\pi)V = 0.637 \ V, \text{ where } V \text{ is the peak value.}$$

5.3.2 The r.m.s. value of a sinusoid

$$v^2 = V^2 \sin^2 \theta$$

$$\text{mean square} = (V^2/\pi) \int_0^\pi \{[1 - \cos(2\theta)]/2\} \, d\theta$$

$$= (V^2/4\pi)[2\theta - \sin(2\theta)]_0^\pi$$
$$= (V^2/4\pi)[2\pi]$$
$$= \tfrac{1}{2}V^2$$

Therefore
$$V_{rms} = V/\sqrt{2} = 0.7071 \ V, \text{ where } V \text{ is the peak value.} \quad (5.1)$$

5.3.3 Calculation of power with a sinusoidal excitation

Instantaneous power $= vi$
Let $v = V \cos(\omega t + \phi_1)$ and $i = I \cos(\omega t + \phi_2)$.
The instantaneous power $= (VI/2)[\cos(\phi_1 - \phi_2) + \cos(2\omega t + \phi_1 + \phi_2)]$.
As the mathematical mean of the time-varying cosine term is zero the mean power dissipated is $(VI/2)\cos(\phi_1 - \phi_2)$.
As $V_{rms} = V_{pk}/\sqrt{2}$ and $\phi_1 - \phi_2$ is the angle between the voltage and

current which may be called ϕ, then the mean power dissipated may be stated as

$$V_{rms} I_{rms} \cos \phi \tag{5.2}$$

5.4 Power factor

Consider a two-terminal network with a phase angle ϕ between the current and voltage as shown in Figure 5.4.

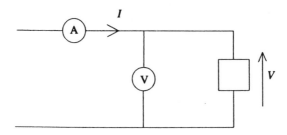

Figure 5.4 Illustrating power factor and volt–ampere product

Ideal meters connected as shown will indicate an r.m.s. value I for the current and an r.m.s. value V for the voltage. The power dissipated is $VI \cos \phi$. However, a voltage source of V volts is delivering a current of I amps. We speak of VI as the volt–ampere product for the network. This is used in power engineering and is also referred to as the *apparent power*.

$$\text{Actual power} = \text{apparent power} \times \cos \phi$$

Hence $\cos \phi$ is known as the *power factor* of the network.

If the positive-going zero crossing of the current sinusoid occurs before that of the voltage sinusoid the current is said to *lead* the voltage or the voltage to *lag* the current. If I leads V we have a leading power factor. If I lags V we have a lagging power factor.

For a resistance the current and voltage are in phase and the power factor is unity. For inductance and capacitance the phase angle between the current and voltage is 90°, and hence the power factor is zero. This confirms that the mean power dissipated in an ideal inductance or capacitance is zero.

5.5 Phasors

In applying Kirchhoff's laws to a.c. voltages and currents the problem arises of adding together, say, two voltages. The r.m.s. values cannot be added arithmetically as the phase angle between the two voltages may not be zero. Consider adding $v_1 = V_{pk1} \sin \omega t$ and $v_2 = V_{pk2} \sin(\omega t + \phi)$.

The algebraic sum of v_1 and v_2 is given by

$$(V_{pk1}^2 + V_{pk2}^2 + 2V_{pk1}V_{pk2} \cos \phi)^{\frac{1}{2}} \sin\{\omega t + \tan^{-1}[V_{pk2} \sin \phi/(V_{pk1} + V_{pk2} \cos \phi)]\}$$

which is not a manageable expression.

The answer to this problem is to use a phasor to represent the magnitude and phase angle of each voltage. A phasor is simply a line on a diagram whose length is proportional to the magnitude of the current or voltage which it represents. The r.m.s. magnitude is normally used. The direction of the line represents the phase angle of the current or voltage. It must be realized that phase only has significance as a phase difference with respect to another phasor or to some arbitrary reference. This is demonstrated in Figure 5.5 where two phasors are drawn to represent v_1 and v_2. In this book bold face type is used to denote a phasor quantity. If v_1 is taken as the reference phase, draw a line horizontally to the right and of length $|\mathbf{V}_1| = V_{pk1}/\sqrt{2}$ units. Then draw a second line of length $|\mathbf{V}_2| = V_{pk2}/\sqrt{2}$ units at an angle ϕ to the horizontal.

In Figure 5.5 phasor $\mathbf{V}_2$ is said to be *leading* phasor $\mathbf{V}_1$, and conversely $\mathbf{V}_1$ is said to be *lagging* phasor $\mathbf{V}_2$. The sum of the two phasors $\mathbf{V}_1$ and $\mathbf{V}_2$ is obtained by completing the parallelogram. Again the reader may show that the magnitude and direction of $\mathbf{V}_{sum}$ corresponds to the algebraic expression given previously. Section 5.7 deals with the use of complex numbers in a.c. analysis and a phasor can be regarded as a diagrammatic way of representing a complex number.

Phasors are very similar to vectors and may be added and subtracted like vectors. Phasors will now be applied to a range of simple circuits.

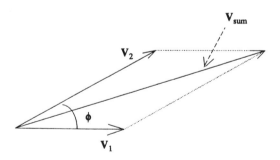

Figure 5.5   A phasor diagram

5.5.1 Inductance

Let $i = I_{pk} \sin \omega t$.

Now

$$v = L \, di/dt$$
$$= LI_{pk}\omega \cos \omega t$$
$$= \omega LI_{pk} \sin(\omega t + \pi/2)$$
$$= V_{pk} \sin(\omega t + \pi/2)$$

Hence $|v| = \omega L|i|$ and v leads i by 90°.

This is represented in the phasor diagram for the inductance shown in Figure 5.6.

ωL is called the *reactance*, X_L, of the inductance.

The following statements are correct:

$$|\mathbf{V}| = \omega L |\mathbf{I}| \quad \text{and} \quad \mathbf{V} \text{ leads } \mathbf{I} \text{ by } 90°$$
$$V = \omega L I$$

where the upper case denotes r.m.s. values.

The equation $V = \omega L I$, or $V = X_L I$, is useful because in many practical cases it is only the r.m.s. magnitudes that are of interest. However, phase relationships cannot be neglected when combining quantities, even when it is only the magnitude of the final answer that is required.

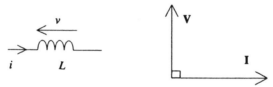

Figure 5.6 Phasor diagram for an inductance

5.5.2 Capacitance

Let $v = V_{pk} \sin \omega t$.
Now

$$i = C \, dv/dt$$
$$= CV_{pk}\omega \cos \omega t$$
$$= \omega C V_{pk} \sin(\omega t + \pi/2)$$

Hence $|i| = \omega C|v|$ or $|v| = |i|/\omega C$ and v lags i by 90°.

The phasor diagram is shown in Figure 5.7.

As the voltage phasor is now pointing downwards the capacitive reactance, X_C, is conventionally taken as negative:

$$X_C = -1/\omega C$$
$$V_{rms} = I_{rms}/\omega C$$

(Note that the r.m.s. value is a magnitude and a sign is not required.)

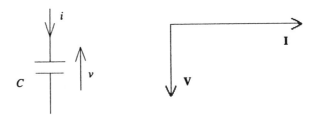

Figure 5.7 Phasor diagram for a capacitance

5.5.3 Inductance and capacitance in series

Note that the phasors V_C and V_L in Figure 5.8 are 180° out of phase. Therefore the magnitude of the total voltage across the circuit, V, is $V_L - V_C$. $V_L = IX_L$ and V_L leads I by 90° and $V_C = IX_C$ and V_C lags I by 90°. The total voltage $V = I(X_L - X_C)$. The total reactance is thus $X_L - X_C$ or $\omega L - 1/\omega C$.

The total reactance of two reactances in series is thus the *sum* of the separate reactances.

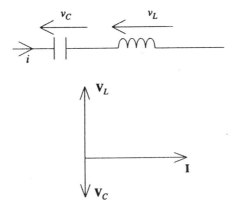

Figure 5.8 Phasor diagram for the series *LC* circuit

5.5.4 Resistance and inductance in series

A phasor diagram for the circuit of Figure 5.9 is shown in Figure 5.10.

The total phasor voltage is $V = (V_R^2 + V_L^2)^{\frac{1}{2}} \angle \phi$, where $\tan \phi = V_L/V_R = \omega L/R$.

Therefore

$$|V| = (I^2 R^2 + \omega^2 L^2 I^2)^{\frac{1}{2}}$$
$$= |I|(R^2 + \omega^2 L^2)^{\frac{1}{2}}$$

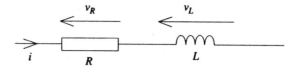

Figure 5.9 A series *RL* circuit

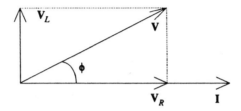

Figure 5.10 Phasor diagram for the series *RL* circuit

When a circuit contains resistive and reactive components we speak of the total *impedance*, Z, of the circuit where $Z = v/i$.

For the case of resistance and inductance in series the magnitude of the impedance is given by

$$|\mathbf{Z}| = (R^2 + \omega^2 L^2)^{\frac{1}{2}}$$

5.5.5 Resistance and capacitance in series

A phasor diagram for the circuit of Figure 5.11 is shown in Figure 5.12. We now have

$$\begin{aligned}|\mathbf{V}| &= (V_R^2 + V_C^2)^{\frac{1}{2}} \\ &= |\mathbf{I}|(R^2 + 1/\omega^2 C^2)^{\frac{1}{2}} \\ \tan\phi &= V_C/V_R = 1/\omega CR\end{aligned}$$

and

$$|\mathbf{Z}| = (R^2 + 1/\omega^2 C^2)^{\frac{1}{2}}$$

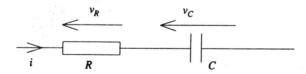

Figure 5.11 A series *RC* circuit

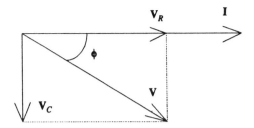

Figure 5.12   Phasor diagram for the series *RC* circuit

It may be concluded that for inductive or capacitive reactances the magnitude of the series circuit impedance is given by the square root of the sum of the resistance squared plus the total reactance squared:

$$|\text{Impedance}| = [(\text{resistance})^2 + (\text{total reactance})^2]^{\frac{1}{2}} \qquad (5.3)$$

_____ **Example 5.2** _____

A capacitor of reactance 20 ohms is connected in series with an inductor across an ideal sinusoidal voltage source. The voltages measured across the capacitor and the inductor are 2 and 7.2 respectively. The d.c. resistance of the inductor is measured as 60 ohms.

Calculate the magnitude of the supply voltage and the inductive reactance of the inductor.

It is always desirable to draw a circuit diagram and mark on it the significant voltages and currents. This is done in Figure 5.13.

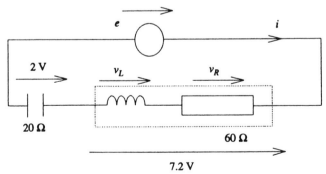

Figure 5.13   Circuit for example 5.2

The dotted line encloses the inductor which is modelled by an inductance in series with a resistance of 60 ohms. Now build up a phasor diagram as shown in Figure 5.14 using the circuit current **I** as the reference phasor. The phasor for the capacitor voltage is known completely and as we also know the capacitive reactance the circuit current may be calculated as 0.1 ampere. This allows the V_R phasor to be drawn. The V_L phasor must lead the current by 90° but is of unknown magnitude as yet. The phasor representing the 7.2 volts across the inductor can now be added.

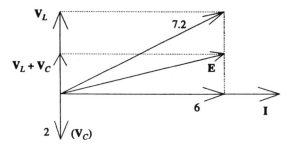

Figure 5.14 Phasor diagram for example 5.2

V_L may be calculated from Pythagoras's theorem:

$$V_L = [(7.2)^2 - (6)^2]^{1/2} = 3.98$$

Hence the inductive reactance is $3.98/0.1 = 39.8$ ohms. Also

$$|V_L + V_C| = 3.98 - 2 = 1.98$$

Whence

$$|E| = [(1.98)^2 + (6)^2]^{1/2} = 6.318 \text{ V}$$

5.5.6 Phasors applied to parallel circuits

The techniques used for series circuits apply to parallel circuits except that Kirchhoff's current law is now required rather than the voltage law.

Consider the circuit of Figure 5.15.

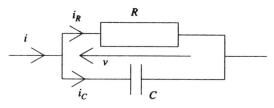

Figure 5.15 A parallel *RC* circuit

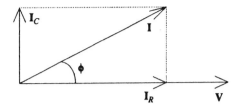

Figure 5.16   Phasor diagram for the parallel *RC* circuit

It is now the voltage which is common to both components and therefore it is convenient to use this as the reference phasor. The phasor diagram is developed in Figure 5.16.

Here

$$|\mathbf{I}_C| = |\mathbf{V}|\omega C \quad \text{and} \quad |\mathbf{I}_R| = |\mathbf{V}|/R$$

also

$$|\mathbf{I}| = (V^2/R^2 + V^2\omega^2C^2)^{\frac{1}{2}}$$
$$= |\mathbf{V}|(1/R^2 + 1/X_C^2)^{\frac{1}{2}} \tag{5.4}$$

The reader should now be able to apply phasors to the analysis of other simple parallel circuits and to more complex circuits involving series and parallel combinations.

5.6 Impedance and admittance

The impedance, Z, of a circuit has already been defined by $Z = v/i$ and it has been shown that for the series circuit

$$|\text{Impedance}| = [(\text{resistance})^2 + (\text{total reactance})^2]^{\frac{1}{2}}$$

Equation (5.4) is in the form of Ohm's law written as $I = V \times$ constant. For the resistive case the constant is called the conductance, G, where $G = 1/R$. For the general a.c. case the constant is called the *admittance*, Y, of the circuit where $Y = 1/Z$.

Whereas the impedance is made up of a resistance and a reactance, the admittance consists of a conductance and a *susceptance*, B. The magnitude of the admittance is given by

$$|\text{Admittance}| = [(\text{conductance})^2 + (\text{susceptance})^2]^{\frac{1}{2}} \tag{5.5}$$

5.7 Symbolic or 'j' notation

Phasor diagrams provide a method of dealing with a.c. circuits which gives a good appreciation of the electrical concepts involved. However, in a solution to a particular problem much trigonometry can be involved. An algebraic solution would be desirable. Chapter 3 introduces the fundamental relationships between voltage and current for passive components. For circuits containing inductance or capacitance or both it can be seen that these expressions lead inevitably to an integro-differential equation. Whilst this is a necessary approach when dealing with non-sinusoidal waveforms it will be shown that the technique generally known as the symbolic or 'j' notation provides a much simpler algebraic solution for circuits involving sinusoidal excitation.

5.7.1 Complex exponential excitation

It is possible to introduce 'j' notation in a very pragmatic way which allows the technique to be used but offers little understanding of its origin or validity. A brief version of a fundamental approach is offered and the reader is referred to texts such as Nilsson (1993) for a fuller treatment.

Consider a current given by $i = Ie^{(j\omega t + \phi)}$. From the Euler identity this may be expressed as

$$i = I\cos(\omega t + \phi) + jI\sin(\omega t + \phi)$$

Thus the real part of this exponential function for current corresponds to a cosinusoidal current. Now let this exponential current flow in a series *LCR* circuit. The impedance of the circuit is given by v/i and hence may be found if the voltage, v, across the circuit can be calculated. From Kirchhoff's voltage law the instantaneous value of v is equal to the sum of the instantaneous values of voltage across each element. Thus

$$v = Ri + L\,di/dt + (1/C)\int i\,dt$$
$$= RIe^{(j\omega t + \phi)} + LIj\omega e^{(j\omega t + \phi)} + (I/Cj\omega)e^{(j\omega t + \phi)}$$
$$= Ie^{(j\omega t + \phi)}(R + j\omega L + 1/j\omega C)$$

Thus the impedance of the circuit for this exponential excitation is

$$R + j\omega L + 1/j\omega C$$

and as the real part of i and of v correspond to the cosinusoidal excitation and response this expression for impedance may be used directly with a cosinusoidal excitation.

It is perhaps appropriate at this point to remind the reader that phasor diagrams and 'j' notation provide analysis techniques for circuits with a sinusoidal, or cosinusoidal, excitation. They have no meaning for non-sinusoidal problems. All the circuit laws, theorems, and analysis techniques developed for d.c. may be applied to a.c. sinusoidal circuits provided jωL is used as the impedance of an inductance and 1/jωC is used for a capacitance. Resistance remains as R.

5.7.2 An alternative approach to 'j' notation

It is possible to arrive at the same result using the following argument. By reference to section 2.1.1 it is seen that 'j' may equally well be regarded as an operator which produces an anticlockwise rotation of 90°. Now consider by way of example an ideal inductance. The magnitude of the impedance of the inductance has been shown to be ωL and the phase angle between the current and voltage is 90° with the voltage leading the current. If the resulting phasor diagram is recognized as being similar to a complex number Argand diagram then the 90° phase difference between the voltage and current is equivalent to multiplying the current by the operator 'j'. Thus the equation

$$v = iZ$$

gives a correct value for v, in phase as well as magnitude, provided Z is taken as jωL. The reader is invited to check this using any simple combination of components to show that the same result is obtained as would be using a phasor diagram approach.

5.8 Basic circuit analysis using 'j' notation

In this section some simple series and parallel circuits are analysed using the imaginary operator j. In each case the components are designated by their complex impedances R, jωL, and 1/jωC and the analysis is done using these values.

5.8.1 Resistance and inductance in series

In Figure 5.17 let **E** be the reference. The circuit impedance is

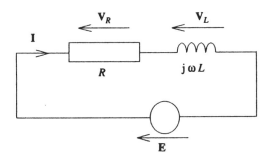

Figure 5.17   A series *RL* circuit

$Z = R + j\omega L$. Therefore the current is $I = E/(R + j\omega L)$, the voltages are $V_R = RI$ and $V_L = j\omega LI$, and the source voltage is $E = V_R + V_L$.

5.8.2 Resistance and capacitance in series

In Figure 5.18 the circuit impedance is $Z = R + 1/j\omega C$. Therefore the current is $I = E/(R + 1/j\omega C)$, the voltages are $V_R = RI$ and $V_C = I/j\omega C$, and the source voltage is $E = V_R + V_C$.

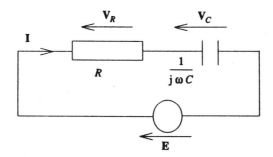

Figure 5.18 A series *RC* circuit

5.8.3 Resistance, inductance and capacitance in series

In Figure 5.19 the circuit impedance is $Z = R + j\omega L + 1/j\omega C$. Therefore the current is $I = E/(R + j\omega L + 1/j\omega C)$, the voltages are $V_R = RI$, $V_L = j\omega LI$, and $V_C = I/j\omega C$, and the source voltage is $E = V_R + V_L + V_C$.

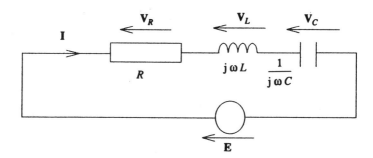

Figure 5.19 A series *RLC* circuit

5.8.4 Resistance and inductance in parallel

In Figure 5.20 let E be the reference. The circuit impedance is $Z = j\omega LR/(R + j\omega L)$. Therefore the current is $I = E(R + j\omega L)/j\omega LR$, the

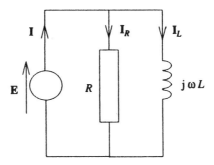

Figure 5.20   A parallel *RL* circuit

other currents are $I_R = E/R$ and $I_L = E/j\omega L$, and the circuit current is $I = I_R + I_L$.

5.8.5 Resistance and capacitance in parallel

In Figure 5.21 let **E** be the reference. The circuit impedance is $Z = R/(j\omega CR + 1)$. Therefore the current is $I = E(j\omega CR + 1)/R$, the other currents are $I_R = E/R$ and $I_C = j\omega CE$, and the circuit current is $I = I_R + I_C$.

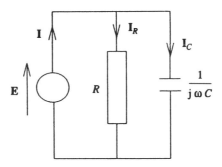

Figure 5.21   A parallel *RC* circuit

5.8.6 Resistance, inductance and capacitance in parallel

In Figure 5.22 let **E** be the reference. The total circuit admittance is $Y = 1/R + 1/j\omega L + j\omega C$.   (The   impedance   is $Z = 1/Y = j\omega LR/[(R - \omega^2 LCR) + j\omega L]$.) The total current is $I = EY = E(1/R + 1/j\omega L + j\omega C)$ and the other currents are $I_R = E/R$, $I_L = E/j\omega L$, and $I_C = j\omega CE$ and $I = I_R + I_L + I_C$.

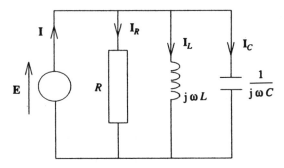

Figure 5.22 A parallel *RLC* circuit

It can be seen that applying 'j' notation leads to equations with *real* and *imaginary* terms. Mathematically this means the quantities are being described using *rectangular coordinates*. Within a calculation it is common practice to use rectangular or polar coordinates, whichever are the more convenient. Additions or subtractions are easy in rectangular form and multiplications or divisions are easier using the polar form. However, the final answer to a problem is best expressed in polar form as in practice a current, voltage, or impedance is normally expressed as a magnitude and a phase angle.

―――― **Example 5.3** ――――――――――――――――――――――――――――

Consider again the circuit of example 5.2.

As before the magnitude of the circuit current, **I**, is calculated as 0.1 ampere and as before this may be taken as the reference phase. Thus $\mathbf{I} = 0.1 + j0$ and $\mathbf{V}_L = j X_L \mathbf{I}$; $\mathbf{V}_R = 60\mathbf{I}$. (**I** is written here with the zero 'j' term to confirm its use as the phase reference and that it has zero phase angle.)

Now $\mathbf{V}_R + \mathbf{V}_L = \mathbf{I}(60 + jX_L)$

and $|\mathbf{V}_R + \mathbf{V}_L| = |\mathbf{I}|[(60)^2 + (XL)^2]^{\frac{1}{2}}$

But this is known to be 7.2 and therefore

$$0.1[(60)^2 + (X_L)^2]^{\frac{1}{2}} = 7.2$$
$$X_L^2 = 72^2 - 60^2 = 1584$$
$$X_L = 39.8 \ \Omega$$

The total voltage is

$$\mathbf{E} = \mathbf{I}(j39.8 - j20 + 60)$$
$$= 0.1(60 + j19.8)$$
$$= 0.1 \times 63.183 \angle 18.263°$$
$$= 6.318 \angle 18.263°$$

_____ **Example 5.4** _____

Calculate the current in the resistance and the voltage across the capacitance for the circuit shown in Figure 5.23.

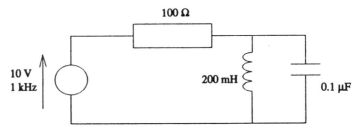

Figure 5.23   Circuit for example 5.4

$$j\omega L = j2\pi 1000 \times 0.2 = j400\pi$$
$$1/j\omega C = 10^6/j2\pi 1000 \times 0.1 = 5000/j\pi = -j5000/\pi$$

The total circuit impedance is

$$Z = 100 + (j400\pi 5000/j\pi)/(j400\pi - j5000/\pi)$$
$$= 100 + 2 \times 10^6/j(-334.91)$$
$$= 100 + j5971.7$$

The current through the resistance is

$$(10 + j0)/(100 + j59) = 1.674 \angle 89.04° \text{ mA}$$

The voltage across the parallel circuit is

$$1.674 \angle 89.04° \times j5971.7$$
$$= 1.674 \angle 89.04° \times 5971.7 \angle 90°$$
$$= 9998 \angle 0.96° \text{ V}$$

5.9 Calculation of power using 'j' notation

It is established in section 5.3.3 that the power dissipated in an impedance is given by $VI \cos \phi$ where V and I are the r.m.s. values of the current through and the voltage across the impedance and ϕ is the angle between them.

Using the 'j' notation the power is equal to the real part of the product of the complex voltage and the complex conjugate of the current. This may be proved as follows.

In the 'j' notation a current $I \angle \phi_1$ is represented by

$$I \cos \phi_1 + jI \sin \phi_1$$

and a voltage $V \angle \phi_2$ by

$$V \cos \phi_2 + jV \sin \phi_2$$

The real part of $(V \cos \phi_2 + jV \sin \phi_2) \times (I \cos \phi_1 - jI \sin \phi_1)$ is

$$VI \cos \phi_1 \cos \phi_2 - jVjI \sin \phi_2 \sin \phi_1 = VI(\cos \phi_1 \cos \phi_2 + \sin \phi_2 \sin \phi_1)$$
$$= VI \cos(\phi_1 - \phi_2)$$

Note that it is equally true to express the power as the real part of the product of the complex current and the complex conjugate of the voltage.

5.10 Maximum power transfer theorem

This theorem has been left until now as it has a.c. aspects that have no equivalent d.c. meaning. The theorem may be stated as follows:

For a given source the maximum power that may be transferred to a load occurs when the load impedance is the complex conjugate of the source impedance. If the load is not freely variable but its magnitude may be varied then a maximum power is obtained in the load when the magnitude of the load impedance is equal to the magnitude of the source impedance.

Consider the circuit shown in Figure 5.24.
Let $\mathbf{Z}_S = R_S + jX_S$ and $\mathbf{Z}_L = R_L + jX_L$. The circuit current is $\mathbf{I} = E/(\mathbf{Z}_S + \mathbf{Z}_L)$. The power, P, in the load equals $R_L|\mathbf{I}|^2 = R_L|E|^2/|[R_S + R_L + j(X_S + X_L)]|^2$.

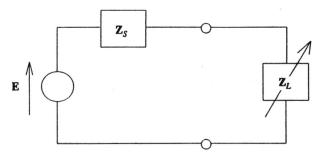

Figure 5.24 Illustrating the maximum power transfer theorem

As the resistances are positive quantities, the maximum value of P occurs when $X_L = -X_S$.

Now

$$P = R_L |\mathbf{E}|^2 / (R_L + R_S)^2$$

$$dP/dR_L = |\mathbf{E}|^2 [1/(R_L + R_S)^2 - 2R_L/(R_L + R_S)^3]$$

For maximum P this must be zero.
Therefore

$$R_L + R_S - 2R_L = 0$$

$$R_L = R_S$$

Thus in total for maximum power

$$\mathbf{Z}_L = R_S - jX_S \tag{5.6}$$

In electronic systems the question frequently arises of 'matching' two parts of the system when they are connected together. It must be realized that there are several rules for matching depending upon the requirements. Within electronic subsystems the important requirement is often for maximum voltage transfer; this produces a different demand on the source and load impedances, or input and output impedances as they are called in that context. In communications matching is needed to minimize signal reflection which leads to the requirement that $\mathbf{Z}_L$ should be equal to $\mathbf{Z}_S$, and not to its conjugate.

When the requirement is for maximum power transfer and the load is freely variable, making the load impedance equal to the complex conjugate of the source impedance is referred to as achieving a conjugate match. It is not uncommon to be able to vary the magnitude, or modulus, of the load but not to alter its phase angle. Making the modulus of the load impedance equal to the modulus of the source impedance is referred to as a modulus match. The power transferred to the load is not as great as would be the case for a conjugate match but it is the maximum that can be obtained given the constraints on the load.

The proof of the modulus condition is left as an exercise for the reader; it will be found easier if the impedances are now represented in polar form rather than in rectangular form, i.e. $\mathbf{Z} = |\mathbf{Z}| \angle \theta°$.

It can be seen that for the d.c. case both conditions reduce to $\mathbf{R}_L = \mathbf{R}_S$.

5.11 Worked examples

Four worked examples are now given on different topics covered in this chapter.

Example 5.5 involves a non-sinusoidal periodic waveform with slope discontinuities. This illustrates the need to use time-dependent expressions in dealing with non-sinusoids and the piecewise approach to problems with waveform slope discontinuities.

Example 5.6 is an illustration of a simple a.c. problem which can be solved without using the 'j' notation or phasor diagrams.

Examples 5.7 and 5.8 demonstrate the use of the 'j' notation.

_____ **Example 5.5** _____

The current shown in Figure 5.25 flows through a series circuit consisting of a 100 Ω resistance, a 100 mH inductance, and a 100 nF capacitance. Calculate expressions for the voltages across each element.

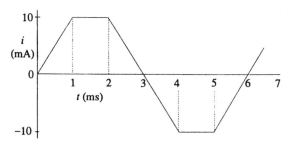

Figure 5.25 Circuit for example 5.5

The voltage across the resistance is given by Ohm's law and therefore has exactly the some waveform as the current with the peak values being plus and minus 1 volt.

The voltage across the inductance is given by $L\, di/dt$. For the time interval (in milliseconds) $0 < t < 1$ the slope of the current waveform is 10 mA/ms which is 10 A/s. The voltage across the inductance is thus a constant 1 volt for this time interval. For $1 < t < 2$ the slope and hence the voltage is zero. For $2 < t < 4$ the slope is -10 and hence the voltage is a constant -1 volt. For $4 < t < 5$ the voltage is again zero and for $5 < t < 7$ it returns to 1 volt.

The voltage across the capacitor is given by

$$v_C = (1/C) \int i\, dt$$

For $0 < t < 1$, $i = 10t$. Therefore

$$v_C = 10^7 \int 10t\, dt = 10^8 t^2/2$$

That is, v_C has a parabolic shape and reaches 50 V at $t = 1$ millisecond.

For $1 < t < 2$ the current into the capacitance is constant and the voltage therefore increases linearly according to the expression

$$v_C = 10^7 \int 10 \times 10^{-3}\, dt = 10^5 t$$

that is, with a slope of 100 V/ms.

For $2 < t < 3$ it is important to realize that the current is still flowing into the capacitance and therefore the voltage continues to rise, although with a continuously reducing slope as the current falls. To calculate an expression for v_C in this interval it is easiest to consider the problem with a time origin starting at $t = 2$. Call this new time variable t'. The equation for i in this interval then becomes

$$i = 10 \times 10^{-3} - 10t'$$

and

$$v_C = 10^7 \int (10^{-2} - 10t')dt' = 10^5 t' - 10^8 t'^2/2$$

Thus in 1 ms of this time the capacitance voltage increases by

$$100 - 100/2 = 50 \text{ volts}$$

These results are combined in the waveform sketch of Figure 5.26.

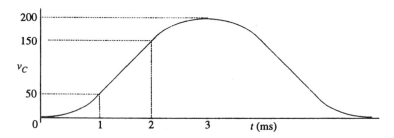

Figure 5.26   Capacitance voltage for example 5.5

The result for the period $3 < t < 6$ is deduced from the symmetry of the problem; however, it could be calculated in the same way as the previous time intervals. It was convenient in places in this problem to work in milliseconds. This avoids the frequent occurrence of 10^{-3} multipliers but care is needed and if in doubt the use of basic units is strongly advised.

_____ **Example 5.6** _____

A simple two-element series circuit takes a current of 100 A when connected to a constant 250 V supply of frequency 50 Hz. The power dissipated in the circuit is 20 kW. It is also known that the current decreases as the frequency increases. Determine the values of the two elements.

To dissipate power one of the elements must be a resistance. As the current decreases, or impedance increases, with increasing frequency the second element must be inductive rather than capacitive.
Power is

$$RI^2 = 20\ 000$$

Therefore

$$R = 20\ 000/10\ 000 = 2\ \Omega$$
$$|Z| = 250/100 = 2.5\ \Omega$$

Therefore

$$(2.5)^2 = \omega^2 L^2 + R^2 = (100\pi)^2 L^2 + 4$$

giving

$$L = 4.78\ \text{mH}$$

_____ **Example 5.7** _____

Calculate the current in the 3 Ω resistance of Figure 5.27.

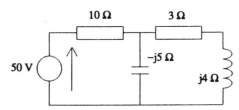

Figure 5.27   Circuit for example 5.7

Solving by mesh analysis label mesh currents I_1 and I_2 to flow clockwise in the left hand and right hand meshes respectively. Applying Kirchhoff's voltage law around each mesh and gathering terms:

$$50 = (10 - j5)I_1 + j5I_2$$
$$0 = j5I_1 + (3 + j4 - j5)I_2$$

From which

$$I_1 = -(3 - j)I_1/j5 = (1 + j3)I_2/5$$

Therefore

$$50 = [5(2 - j)(1 + 3j)/5 + j5]I_2$$
$$I_2 = 50/(2 + 3 - j + 6j + 5j) = 50/(5 + 10j) = 50/11.18 \angle 63.435°$$
$$I_2 = 4.47 \angle -63.435°\ \text{A}$$

Solving by nodal analysis label the voltage at the top of the capacitance as **V**. Applying Kirchhoff's current law at this node and gathering the terms in **V**:

$$\mathbf{V}[1/10 + 1/{-}j5 + 1/(3 + j4)] - 50/10 = 0$$
$$\mathbf{V}[j3 - 4 - 6 - j8 + j10]/[10j(3 + j4)] - 5 = 0$$
$$\mathbf{V} = 50j(3 + j4)/(-10 + j5)$$

Now the current in the 3 Ω resistance is

$$\mathbf{V}/(3 + j4) = 50j/(-10 + j5) = 50 \angle 90°/11.18 \angle 153.435°$$
$$= 4.47 \angle -63.435° \text{ A}$$

_____ **Example 5.8** _____

Referring to the circuit of Figure 5.28 calculate the power dissipated in each resistance and the power delivered by the voltage source.

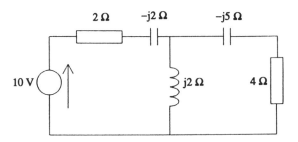

Figure 5.28   Circuit for example 5.8

Label the voltage at the top of the inductance as **V**. Applying Kirchhoff's current law at that node and gathering terms:

$$\mathbf{V}[1/(2 - j2) + 1/j2 + 1/(4 - j5)] - 10/(2 - j2) = 0$$
$$\mathbf{V}[j4 + 5 + 4 - 5 - j9 + j2 + 2]/j2(1 - j)(4 - j5) = 10/(2 - j2)$$
$$\mathbf{V}[6 - j3] = 10j(4 - j5)$$
$$\mathbf{V} = 10j(4 - j5)/3(2 - j)$$

Therefore the current in the 4 Ω branch is

$$10j(4 - j5)/3(2 - j)(4 - j5) = 10j/3(2 - j)$$

The power in the 4 Ω resistance is given by

$$RI^2 = 4 \times 10^2/3^2(2^2 + 1^2) = 8.889 \text{ W}$$

The current in the 2 Ω branch is

$$[10j(4-j5)/3(2-j) - 10]/(2-j2) = [j40 + 50 - 60 + j30]/3(2-j)(2-2j)$$

which simplifies to $(-11+j2)/3$. The power in the 2 Ω resistance is

$$RI^2 = 2(11^2 + 2^2)/9 = 27.778 \text{ W}$$

The power delivered by the source equals the real part of $\mathbf{VI^*}$, where $\mathbf{I^*}$ is the complex conjugate of $\mathbf{I}$:

$$\text{real part of } 10(-11 - j2)/3 = 110/3 = 36.667 \text{ W}$$

Note as a check that $8.889 + 27.778 = 36.667$; that is, the total power dissipated equals the power delivered.

_____ **Problems** _____

P5.1 The repeating sawtooth current function shown in Figure P5.1 flows in a 5 Ω resistance. Plot curves of instantaneous voltage and power as functions of time and determine the value of the average power.

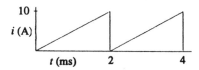

Figure P5.1

P5.2 An ideal 3 mH inductance passes a current with the waveform shown in Figure P5.2. Determine and sketch the voltage $v(t)$ and the instantaneous power $p(t)$. What is the average power, P?

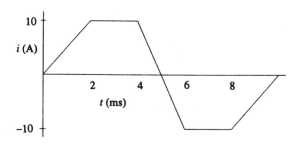

Figure P5.2

P5.3 A series circuit consists of a 2 Ω resistance, a 2 mH inductance, and a 500 μF capacitance. The circuit current has the waveform shown in Figure P5.3. Determine the voltage across each element and sketch each voltage to the same time scale. Also sketch $q(t)$, the charge on the capacitor.

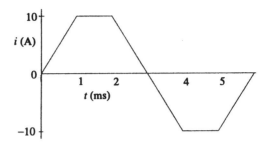

Figure P5.3

P5.4   Determine the average and r.m.s. values of the waveform shown in Figure P5.1.

P5.5   A triangular voltage waveform of period 20 ms has equal positive and negative slopes and peak values of ±5 V. Determine the average and r.m.s. values of the waveform.

P5.6   A full wave rectified sine wave is clipped at 0.707 of its maximum value as shown in Figure P5.4. Determine the average and r.m.s. values of the function.

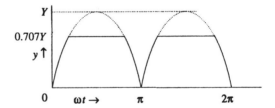

Figure P5.4

P5.7   A voltage source $e(t) = 100 \cos t$ is applied to a series circuit consisting of a 1 Ω resistance and a 1 H inductance. Determine the current, $i(t)$, that flows in the circuit.

P5.8   A series circuit consists of a 1 μF capacitance and a 25 Ω resistance. The circuit is fed from a 100 V, 50 Hz source of zero impedance. Determine the reactance and the impedance of the circuit, the magnitude of the current, and its phase relationship with the applied voltage.

P5.9   The current in a 2 Ω resistor has the waveform shown in Figure P5.5 with a maximum value of 5 A. The average power dissipated in the resistor is 20 W. Determine the angle θ.

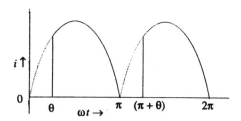

Figure P5.5

P5.10 A parallel circuit consists of an inductor, having an effective series inductance of 2 H and resistance of 20 Ω, and a pure resistance of 100 Ω. The circuit is excited from a 200 V, 100 Hz, zero-impedance source. Determine the magnitude of the current taken from the source and the phase relationship between the current and the applied voltage. Determine also the total power dissipated in the circuit.

P5.11 A voltage source of magnitude 10 volts r.m.s. and angular frequency $\omega = 1000$ rad/s is applied in turn to each of the circuits of Figure P5.6. In each case draw a phasor diagram, calculate the magnitude of the current drawn from the source, and calculate the phase difference between the current and the applied voltage.

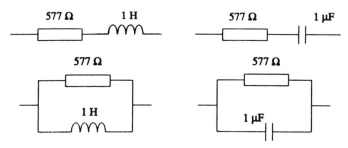

Figure P5.6

P5.12 In a simple two-element series circuit the current lags the applied voltage by 75° at a frequency of 60 Hz. If one of the elements is a 5 ohm resistance determine the second element.

P5.13 A 5 ohm resistance and an unknown capacitance are connected in series. At a frequency of 320 Hz there is a voltage of 25 volts across the resistor. If the current leads the applied voltage by 60° determine the unknown capacitance.

P5.14   A series circuit of inductance 0.05 H and unknown capacitance has the following applied voltage and resulting current:

$$v = 100 \sin(5000t) \quad i = 2 \sin(5000t + 90°)$$

Determine the value of the unknown capacitance.

P5.15   A series *LCR* circuit consists of a 50 mH inductance, a 10 nF capacitance, and a 5 Ω resistance. A sinusoidal voltage of 10 volts is applied to this circuit. Determine the frequency of the supply if the circuit current is in phase with the supply voltage, the value of the circuit current under these conditions, and the voltage across the inductance.

P5.16   A variable frequency sinusoidal voltage is applied across a series circuit consisting of $R = 5\ \Omega$, $L = 0.02$ H, and $C = 80\ \mu$F. Find the values of ω for which the current will (a) lead the voltage by 45°, (b) be in phase, and (c) lag the voltage by 45°.

P5.17   A circuit consists of two parallel branches. The first of these is a series circuit of inductance 0.5 H and resistance 20 Ω. The second is a capacitance of 1 μF. Determine the impedance of this circuit (a) at a frequency of 1 kHz and (b) at the frequency at which it is purely resistive.

P5.18   The circuit constants *R* and *L* of a coil are to be determined by connecting the coil in series with a resistance of 25 Ω and applying a 120 V, 60 Hz source of zero internal impedance across the combination. If the voltage across the resistance is measured as 70.8 V and that across the coil as 86 V determine *R* and *L*.

P5.19   Three impedances, $Z_1 = 5 + j5$ ohms, $Z_2 = j8$ ohms, and $Z_3 = 4$ ohms are connected in series across an unknown voltage source. Determine the value of the voltage source and the current drawn from it if the voltage across Z_3 is $63.2 \angle 18.45°$ volts.

P5.20   In the series circuit shown in Figure P5.7 the voltage across the inductance is $13.04 \angle 15°$ volts. Determine Z.

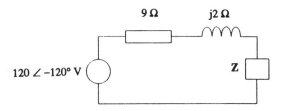

Figure P5.7

P5.21 A series circuit consists of a 1 Ω resistance, a 4 Ω inductive reactance, and an impedance **Z**. If the applied voltage is 50 ∠ 45° V and the resulting current 11.2 ∠ 108.3° A, calculate the impedance **Z**.

P5.22 An 18 Ω resistance is connected in parallel with the series combination of a 3 Ω inductive reactance and a 3 Ω resistance. An ammeter is connected in series with the total circuit. An ideal voltmeter placed across the 3 Ω resistance reads 45 volts. Calculate the reading on the ammeter.

P5.23 For the circuit shown in Figure P5.8, determine the voltage across the 20 Ω resistance using the mesh current method of analysis.

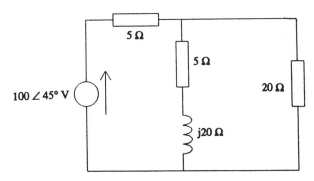

Figure P5.8

P5.24 For the circuit shown in Figure P5.8, determine the voltage across the 20 Ω resistance using the nodal method of analysis.

P5.25 For the circuit of Figure P5.9, determine the power supplied by the source and the power dissipated in each resistor.

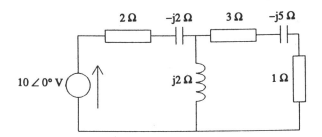

Figure P5.9

P5.26 For the circuit of Figure P5.10, determine the power which each source supplies to the network.

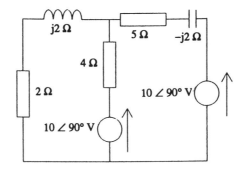

Figure P5.10

P5.27   For the circuit of Figure P5.11 verify that for all values of R and X the r.m.s. value of V_{ab} is 50 V. It will be sufficient to prove this for any arbitrarily chosen values for R and X.

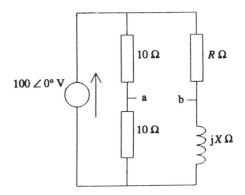

Figure P5.11

CHAPTER SIX

Transient behaviour of circuits

The analysis of circuits considered so far has assumed that the excitation of the circuit has been present for a long time and remains constant. This is referred to as the *steady state* condition of the circuit and the result obtained from the analysis is the *steady state* response. It is sometimes necessary to calculate the response of a circuit to a change in the excitation, e.g. the initial switching on of the circuit. This is called the *transient* response of the circuit. This chapter considers such problems. They are introduced in stages starting with the simplest circuits containing only one type of reactance. It is found that applying basic electrical laws to such circuits results in a differential equation. Circuits containing resistance and either inductance or capacitance, but not both, result in equations where the highest derivative is the first derivative. Such equations are called *first-order* equations and the circuits *first-order* circuits. The equations are solved using classical mathematical methods as these give the best insight into the nature of the circuits and their responses. The reader who is conversant with transform or operator methods of solution is invited first to study the material as it is presented. Once understanding is achieved the most powerful or appropriate mathematical methods may be used to solve the equations resulting from any particular problem.

6.1 First-order circuits with no external excitation

If a circuit has some stored energy but is no longer connected to any external excitation then the behaviour of the circuit currents and voltages is determined by the circuit itself and is independent of any external source. This behaviour is therefore called the *natural response* of the circuit.

In section 6.1.1 the basic electrical laws are applied to an *LR* circuit, resulting in a differential equation. A basic mathematical solution method is then covered in section 6.1.2 and some general results deduced. In section 6.1.3 the differential equation derived in 6.1.1 is solved using the method of 6.1.2. It is desirable that the reader can deduce and solve the equations mathematically when meeting a circuit for the first time and be able to recognize commonly occurring circuits and apply general results without resorting to a complete solution from first principles. This approach is then used for the *CR* circuit.

6.1.1 The *LR* circuit: deriving the equation

Consider the circuit shown in Figure 6.1 with no external excitation for time after $t = 0$. Assume there is some initial energy stored such that the current has an initial value I at time $t = 0$.

Applying Kirchhoff's voltage law gives

$$v_L + v_R = 0$$

i.e.

$$L \, di/dt + Ri = 0 \qquad (6.1)$$

Consider now a solution to equations of this form.

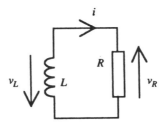

Figure 6.1   The *LR* circuit with no external excitation

6.1.2 First-order equations of the form $dx/dt + ax = 0$

These arise in the analysis of circuits containing only one reactance type and having no external excitation but involving some initial energy storage. The general variable x will be a current or a voltage in an electrical problem. This form of equation is known as a *variables separable* form and it may be rearranged as follows:

$$(1/x)dx = -a \, dt$$

Integrating both sides gives

$$\int (1/x)\mathrm{d}x = -\int a \, \mathrm{d}t$$

$$\ln (x) = -at + k$$
$$x = e^{-at + k} = e^{-at}e^{k}$$

Now if x has some value X_0 at time $t = 0$ then

$$x = X_0 e^{-at} \tag{6.2}$$

This describes the exponential decay of x from some initial value X_0. Examine some features of this by considering a normalized expression

$$y = e^{-\tau} \quad \text{where } y = x/X_0 \text{ and } \tau = at$$

Calculating a few values:

at	x/X_0
0.5	0.6065
1	0.367 88
2	0.135 34
4	0.0183
5	0.006 74

Hence although mathematically x never reaches zero it has fallen to less than 5% of its initial value by the time $t = 3/a$ and is less than 1% by the time $t = 5/a$.

Now calculate the slope of the curve at $t = 0$:

$$\mathrm{d}y/\mathrm{d}\tau = -e^{-\tau}$$

Therefore at $\tau = 0$ the slope is -1. This is illustrated in the plot of Figure 6.2.

The time interval $1/a$ clearly has significance. It is called the *time constant* of the response. The tangent to the curve at $t = 0$ intersects the baseline at a time t equal to one time constant when the curve itself has

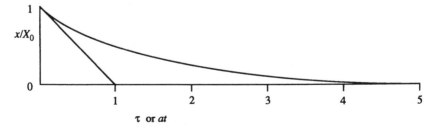

Figure 6.2 Plot of exponential decay

fallen to a value $1/e$ or 0.368. After five time constants the curve has fallen below 1% of its original value and in engineering terms is frequently regarded as negligible after this time.

6.1.3 The *LR* circuit: solving the equation

Rewriting equation (6.1) as

$$di/dt + (R/L)i = 0$$
$$\int (1/i)di = -(R/L) \int dt$$
$$\ln(i) = -(R/L)t + k$$
$$i = e^{-(R/L)t}e^k$$

At $t = 0$ and $i = I_0$ therefore

$$I_0 = e^k$$

hence

$$i = I_0 e^{-(R/L)t} \tag{6.3}$$

Thus the current decays exponentially from its initial value with a time constant of L/R seconds. The value of the time constant will vary considerably depending upon the application.

_____ **Example 6.1** _____

A small relay of inductance 2 H and winding resistance 500 Ω is used with a diode and resistance R to prevent large voltages being developed on switch-off. This is shown in Figure 6.3.

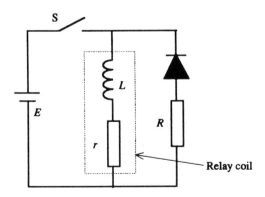

Figure 6.3   Relay circuit for example 6.1

Calculate the release time of the relay in this circuit if it is known that $E = 12$ V, $R = 2$ kΩ, and the relay releases when its current falls below 5 mA.

When the switch, S, has been closed for a considerable time and steady conditions prevail the current through the relay equals E/r or 24 mA. For switch-off take $t = 0$ to be the moment the switch, S, is opened. The e.m.f. induced in the inductance due to the changing current is such as to oppose the change and the diode now conducts. Neglecting the forward resistance of the diode the total circuit resistance is now $(r + R)$ or 2300 Ω.

Invoking equation (6.3), the current in the relay is therefore

$$24e^{-(2300/2)t} \text{ mA}$$

The relay releases when the current is 5 mA. Therefore

$$5 = 24e^{-(2300/2)t}$$

giving

$$t = 1.36 \text{ ms}$$

6.1.4 The CR circuit

Consider the circuit shown in Figure 6.4 and assume the capacitor voltage has an initial value V_0 volts.

For the circuit as labelled with the current, i, flowing out of the capacitance at the positive end of v_C:

$$v_C = -(1/C) \int i \, dt \quad \text{and} \quad v_R = iR$$

Applying Kirchhoff's voltage law

$$v_C = v_R$$

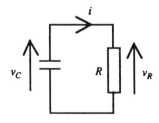

Figure 6.4   The CR circuit with no external excitation

Therefore

$$iR + (1/C) \int i \, dt = 0$$

$$di/dt + (1/CR)i = 0$$

$$\int (1/i)di = -(1/CR) \int dt$$

$$\ln(i) = -t/CR + k$$

$$i = e^{-t/CR}e^{k}$$

But at $t = 0$, $v_C = V_0$ and therefore $i = V_0/R$. Hence

$$V_0/R = e^{k} \quad \text{or} \quad i = (V_0/R)e^{-t/CR}$$

and

$$v_C = V_0 e^{-t/CR} \tag{6.4}$$

6.2 First-order circuits with step function excitation

The step function is a simple, but commonly occurring, circuit excitation. Using the CR circuit as an example the following sections derive the differential equations, offer a solution method, and make some general deductions, in the same way as the natural responses are treated in the previous sections.

6.2.1 The CR circuit with no initial stored energy

Consider the circuit shown in Figure 6.5. If the switch is closed at $t = 0$ the voltage applied across the circuit is zero for all time before $t = 0$ and is equal to E volts for all positive values of t. This is called a step function of value E.

There are two approaches to a first-order circuit with a step excitation. Applying Kirchhoff's voltage law produces a slightly different form of

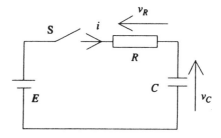

Figure 6.5 *CR* circuit with step function excitation

equation from that produced by his current law. Both solutions are given in this case and the reader is left to consider the relative merits of each for future use.

With the switch closed, applying the voltage law gives

$$E = iR + (1/C) \int i \, dt$$

Differentiating:

$$0 = di/dt + i/CR$$

$$\int (1/i)di = -(1/CR) \int dt$$

$$\ln(i) = -t/CR + k$$

$$i = e^{-t/CR}e^k$$

At $t = 0$, $i = E/R$ or $E/R = e^k$. Therefore

$$i = (E/R)e^{-t/CR} \qquad (6.5)$$

The voltage across the capacitance may then be found if required:

$$v_C = (1/C) \int i \, dt = (E/CR) \int e^{-t/CR} \, dt = -E \, e^{-t/CR} + k$$

But at $t = 0$, $v_C = 0$ or $k = E$. Therefore

$$v_C = E(1 - e^{-t/CR}) \qquad (6.6)$$

This equation is plotted in Figure 6.6. It should be compared with the exponential decay described by equation (6.2) which is plotted in Figure 6.2. Applying the same considerations to the curve of Figure 6.6 it can be deduced that it takes infinite time for v_C to reach the value E but that

> v_C reaches 63% of E after one time constant of the circuit,
> v_C reaches 95% of E after three time constants,
> v_C reaches 99% of E after five time constants.

It can also be seen that the asymptote to the curve at $t = 0$ reaches a value E after a time equal to one time constant (τ).

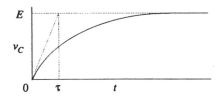

Figure 6.6 A plot of equation (6.6)

An alternative approach to the analysis of a first-order circuit with a step function excitation is to apply Kirchhoff's current law. Doing this at the junction of the resistance and the capacitance in the circuit of Figure 6.5 gives

$$(E - v_C)/R = C \, dv_C/dt$$

or

$$dv_C/dt + (1/CR)v_C = E/CR \tag{6.7}$$

This is a similar form of differential equation to that encountered previously but the right hand side is now a constant rather than zero. A solution method is suggested in section 6.2.2.

6.2.2 First-order equations of the form dx/dt + ax = a constant

These arise from circuits containing resistance and one reactance type with a constant or step function excitation:

$$dx/dt + ax = k$$

Multiply both sides by e^{at}:

$$e^{at} \, dx/dt + e^{at} ax = ke^{at}$$

$$\frac{d}{dt}(xe^{at}) = ke^{at}$$

$$x \, e^{at} = \int k \, e^{at} \, dt = (k \, e^{at}/a) + c$$

$$x = k/a + ce^{-at}$$

Now if $x = X_0$ at time $t = 0$ then $c = X_0 - k/a$ giving

$$x = k/a(1 - e^{-at}) + X_0 e^{-at} \tag{6.8}$$

The solution is seen to consist of two terms. The term $(X_0 - k/a)e^{-at}$ is called the transient response and the term k/a is the steady state response.

6.2.3 The CR circuit: an alternative solution

Apply the method developed in section 6.2.2 to equation (6.7) deduced in 6.2.1:

$$dv_C/dt + (1/CR)v_C = E/CR$$

Multiply by $e^{t/CR}$:

$$e^{t/CR}\,dv_C/dt + e^{t/CR}(1/CR)v_C = (E/CR)e^{t/CR}$$

$$d(v_C e^{t/CR})/dt = (E/CR)e^{t/CR}$$

$$v_C\,e^{t/CR} = (E/CR)\int e^{t/CR} = E\,e^{t/CR} + k$$

$$v_C = E + ke^{-t/CR}$$

Now if $v_C = 0$ at $t = 0$ then $k = -E$ and

$$v_C = E(1 - e^{-t/CR}) \tag{6.9}$$

With either solution if the circuit has some stored energy at time $t = 0$ then the initial conditions will be different but the solution process is unchanged. Care is sometimes needed in translating the initial state of the circuit into initial values for either current or voltage.

6.2.4 Examples

The following worked examples demonstrate that although the student should be able to set up and solve the electrical equation for any first-order circuit, it is frequently possible to recognize the circuit as being equivalent to those for which general solutions have been deduced. These solutions may then be applied using appropriate values.

Example 6.2

In the circuit shown in Figure 6.7 the switch is closed at time zero and is opened again after 10 ms. Deduce expressions to describe the variation of v_C with time. Assume the capacitance is initially uncharged.

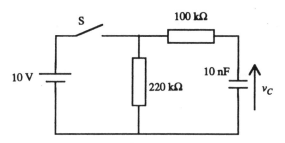

Figure 6.7 Circuit for example 6.2

When the switch is closed the 220 kΩ resistance appears across the battery and has no effect on the charging of the capacitance. An equation for v_C may thus be written as

$$v_C = 10(1 - e^{-t/(1E-8)(1E-5)})$$
$$= 10(1 - e^{-1000t})$$
$$= 10(1 - e^{-t})$$

if t is expressed in milliseconds. This expression is valid over the time range $0 < t < 10$ ms.

The time constant is 1 ms and therefore the capacitance voltage may be assumed to have reached 10 V after 10 ms (it is accurately 9.999 55 V).

After the switch is reopened the capacitance discharges through the two resistances in series. It is easy to write an equation for v_C if a new origin for time is taken at the moment the switch opens. However, confusion can be caused if this equation is used together with the equation for the first time period. A proper way to proceed is to use a new time variable t' where $t' = t - 10$ (still working in milliseconds). The equation for v_C is then

$$v_C = 10e^{-t'/(1E-8)(3.2E5)(1000)}$$

where the 1000 is required if t' is to be in milliseconds. This gives

$$v_C = 10e^{-0.3125t'}$$

This is a convenient form for sketching v_C as the differing time variables are easily handled. However, if any algebraic manipulation is required t' must be replaced.

$$v_C = 10e^{-0.3125t'} = 10e^{-0.3125(t-10)} = 10e^{-0.3125t}e^{3.125}$$

or

$$v_C = 227.6e^{-0.3125t}$$

Example 6.3

For the circuit shown in Figure 6.8 deduce an expression for the current, i, in the inductance. The switch, S, is closed at time zero and there is no initial energy stored in the inductance.

Let the current through the 220 Ω resistance be i'. Applying Kirchhoff's voltage law around the right hand loop of the circuit:

$$100i + 1\, di/dt = 220i'$$

and around the outer loop:

$$220i' + 470(i + i') = 10$$

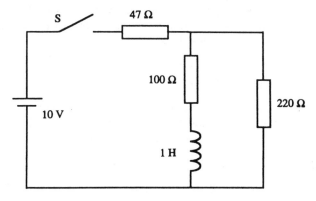

Figure 6.8   Circuit for example 6.3

Therefore

$$i' = (10 - 47i)/267$$

which gives

$$100i + \mathrm{d}i/\mathrm{d}t = 220 \times 10/267 - 220 \times 10 \times i/267$$
$$\mathrm{d}i/[138.73(i - 0.0594)] = -\mathrm{d}t$$

Integrating both sides

$$\ln(i - 0.0594) = -138.73t + c$$

or

$$i - 0.0594 = \mathrm{e}^{-138.73t}\mathrm{e}^{c}$$

At $t = 0$, $i = 0$. Therefore

$$\mathrm{e}^{c} = -0.0594$$
$$i = 0.0594(1 - \mathrm{e}^{-138.73t})$$

That is, the current in the inductance reaches a final value of 59.4 mA with a time constant of 1/138.73 or 7.21 ms. The final value may be checked by shorting out the inductance and calculating the d.c. current in the 100 Ω resistance.

6.3 Second-order circuits

These are circuits containing resistance and both types of reactance. Applying the fundamental electrical laws results in a second-order differential equation for both current and voltage. Once more the approach adopted is to derive the equation, offer a mathematical solution method which enables some general results to be considered, and finally apply these techniques to example circuits.

6.3.1 The series *LCR* circuit

Consider by way of example the series *LCR* circuit shown in Figure 6.9.

Assume that at time $t = 0$ the current $i = I$ and the capacitance voltage $v_C = V$. Applying Kirchhoff's voltage law:

$$v_L + v_R + v_C = 0$$

$$L \, di/dt + Ri + (1/C) \int i \, dt = 0$$

Differentiating:

$$L \, d^2i/dt^2 + R \, di/dt + i/C = 0 \tag{6.10}$$

A solution method is now required for this second-order differential equation.

6.3.2 Second-order differential equations

Consider an equation of the general form

$$d^2x/dt^2 + 2a \, dx/dt + \omega^2 x = 0 \tag{6.11}$$

It will become clear later why the coefficients $2a$ and ω^2 are expressed in this way.

Try a solution $x = Ae^{mt}$. Substituting this into the general equation gives

$$m^2 Ae^{mt} + 2amAe^{mt} + \omega^2 Ae^{mt} = 0$$

or

$$Ae^{mt}(m^2 + 2am + \omega^2) = 0$$

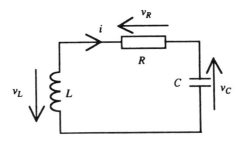

Figure 6.9   Series *LCR* circuit with initial energy storage

Thus $Ae^{mt} = 0$, which is not a useful solution, or

$$(m^2 + 2am + \omega^2) = 0$$

This last equation is called the *characteristic* or *auxiliary* equation relating to equation (6.11).

Ae^{mt} is a solution if $(m^2 + 2am + \omega^2) = 0$ which gives

$$m = -a \pm (4a^2 - 4\omega^2)^{1/2}/2$$
$$= -a \pm (a^2 - \omega^2)^{1/2}$$

Therefore the complete solution to the differential equation is

$$x = A_1 e^{(-a + \sqrt{a^2 - \omega^2})t} + A_2 e^{(-a - \sqrt{a^2 - \omega^2})t} \qquad (6.12)$$

The constants A_1 and A_2 are found by substituting the initial conditions existing in a particular problem.

The form of the solution depends upon the relative values of a and ω and Figure 6.10 illustrates the general characteristics of the three possible cases.

1. $a > \omega$: In this case the shape of the solution is characterized by an initial surge followed by an exponential decay.

   This is known as the *overdamped* case.
2. $a < \omega$: The shape of the solution is now a sinusoid with an exponentially decreasing amplitude.

   This is known as the *underdamped* case.

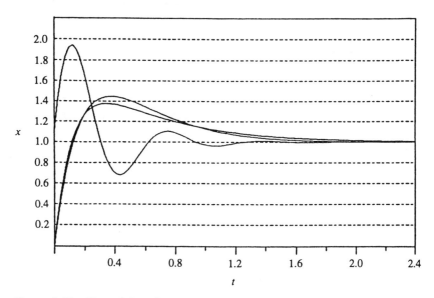

Figure 6.10 Plots of the solutions to equation (6.12)

3. $a = \omega$: The shape is similar to case 1 but the solution reaches its final steady value in the minimum time possible.

This is known as the *critically* damped case.

It can be seen from equation (6.12) that in this case the solution appears to be trivial. A further step is required for a solution as will be shown later.

6.3.3 Solving the series *LCR* circuit

Rearranging equation (6.10) gives

$$\mathrm{d}^2 i/\mathrm{d}t^2 + (R/L)\mathrm{d}i/\mathrm{d}t + i/LC = 0 \qquad (6.13)$$

Substituting a trial solution $i = Ae^{mt}$ leads to the characteristic equation:

$$m^2 + (R/L)m + 1/LC = 0 \qquad (6.14)$$

or

$$m = -R/2L \pm [(R/L)^2 - 4/LC]^{\frac{1}{2}}/2$$

Thus the form of the solution depends on the component values:

If $R^2 > 4L/C$ the response is overdamped.
If $R^2 = 4L/C$ the response is critically damped.
If $R^2 < 4L/C$ the response is underdamped.

Note that the underdamped case involves the lowest values for circuit resistance. This corresponds to the high Q circuit in the steady state frequency domain (see section 7.1.2 for an introduction to the Q factor).

These results justify pursuing this problem in terms of component parameters R, L, and C. However, to proceed further leads to rather tedious algebraic expressions which make it difficult to see the significance of the results. Hence at this stage values will be assigned to the components and the value of R will be adjusted to give each of the three cases in turn.

6.3.4 The overdamped series *LCR* circuit

Take $L = 1$ H, $C = 0.0625$ F, and $R = 10 \, \Omega$ and let the initial current be 1 A and the initial capacitance voltage be 4 V.

Substituting these values in equation (6.14) gives

$$m^2 + 10m + 16 = 0$$
$$m = -10/2 \pm (100 - 64)^{\frac{1}{2}}/2 = -5 \pm 6/2 = -8 \text{ or } -2$$

Therefore the solution is

$$i = A_1 e^{-8t} + A_2 e^{-2t}$$

The values of A_1 and A_2 can always be found by calculating the initial value of the main variable, in this case the current, i, and the initial value of its first derivative. These are then equated to the given initial electrical conditions for the circuit.

Take i first. At $t=0$, $i=A_1+A_2$; but the initial current is given as 1 A and therefore

$$A_1 + A_2 = 1 \tag{6.15}$$

Now take di/dt:

$$di/dt = -8A_1e^{-8t} - 2A_2e^{-2t} = -8A_1 - 2A_2 \quad \text{at } t=0$$

Referring back to the circuit (Figure 6.5) Kirchhoff's voltage law gives

$$L\,di/dt + Ri + v_C = 0$$

Substituting values for R, L, and the initial values of i and v_C gives, at $t=0$,

$$di/dt = -10 - 4 = -14$$

Therefore

$$-8A_1 - 2A_2 = -14 \tag{6.16}$$

Solving the two equations (6.15) and (6.16) in A_1 and A_2 gives

$$A_1 = 2 \quad \text{and} \quad A_2 = -1$$

giving a final solution of

$$i = 2e^{-8t} - e^{-2t} \tag{6.17}$$

A plot of this solution is shown in Figure 6.11. A plot of the solution to the critically damped case is also shown on the same axes for comparison.

6.3.5 The underdamped *LCR* circuit

Consider the same circuit with the same values except for R which is now given the value $4\sqrt{3}$ ohms. This odd value is chosen simply to give integer numbers in demonstrating the solution. The characteristic equation in this case becomes

$$m^2 + 4\sqrt{3}m + 16 = 0$$
$$m = -2\sqrt{3} \pm (48 - 64)^{\frac{1}{2}}/2$$
$$= -2\sqrt{3} \pm j2$$

Therefore

$$i = A_1e^{(-2\sqrt{3}+j2)t} + A_2e^{(-2\sqrt{3}-j2)t}$$

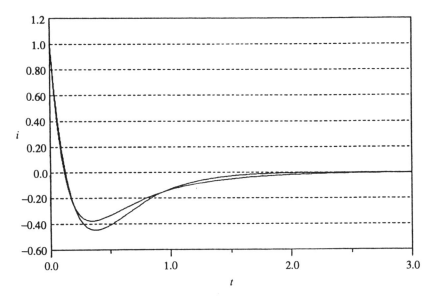

Figure 6.11   Plots of overdamped and critically damped solutions

Rearranging:

$$i = e^{-2\sqrt{3}t}(A_1 e^{j2t} + A_2 e^{-j2t})$$
$$= e^{-2\sqrt{3}t}[A_1(\cos 2t + j \sin 2t) + A_2(\cos 2t - j \sin 2t)]$$
$$= e^{-2\sqrt{3}t}[(A_1 + A_2)\cos 2t + j(A_1 - A_2)\sin 2t]$$
$$= e^{-2\sqrt{3}t}(B \cos 2t + jD \sin 2t)$$

where B and D are constants.

Thus the response is a sinusoid of angular frequency 2 and of exponentially decaying amplitude. The process to evaluate the constants is the same in all cases. At $t = 0$, $i = B$, but the initial value of current is 1 A and therefore $B = 1$. Differentiating i with respect to t and substituting $t = 0$ gives

$$di/dt = j2D - 2\sqrt{3}$$

But from the circuit

$$L \, di/dt = -v_C - Ri$$

which at $t = 0$ becomes

$$di/dt = -4 - 4\sqrt{3}$$

Therefore

$$j2D - 2\sqrt{3} = -4 - 4\sqrt{3}$$
$$D = j(2 + \sqrt{3})$$

The complete solution is

$$i = e^{-2\sqrt{3}t}[\cos 2t + j(2 + \sqrt{3})\sin 2t] \tag{6.18}$$

A plot of this solution is shown in Figure 6.12.

6.3.6 The critically damped *LCR* circuit

Consider the same circuit with the same values except for R which is now given the value 8 ohms. The characteristic equation in this case becomes

$$m^2 + 8m + 16 = 0$$
$$m = -8/2 \pm (64 - 64)^{\frac{1}{2}}/2 = -4$$

Thus apparently $i = Ae^{-4t}$, but for a second-order equation two constants of integration are required. Try an additional solution $i = Bte^{-4t}$.

For this solution

$$di/dt = -4Bte^{-4t} + Be^{-4t}$$
$$d^2i/dt^2 = 16Bte^{-4t} - 4Be^{-4t} - 4Be^{-4t}$$

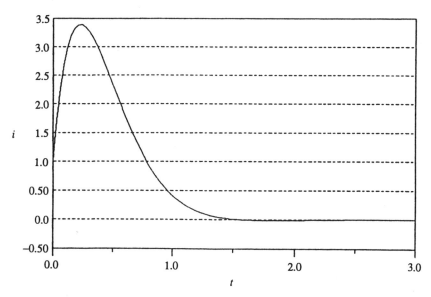

Figure 6.12 A plot of the underdamped solution (equation (6.18))

Substituting into equation (6.13) gives

$$16Bte^{-4t} - 4Be^{-4t} - 4Be^{-4t} + 8(-4Bte^{-4t} + Be^{-4t}) + 16Bte^{-4t} = 0$$

which is correct; hence $i = Bte^{-4t}$ is a solution.
    The full solution is thus $i = Ae^{-4t} + Bte^{-4t}$.
    At $t = 0$, $i = A$ and therefore $A = 1$ and

$$di/dt = -4Ae^{-4t} - 4Bte^{-4t} + Be^{-4t} = -4 + B$$

But from the circuit

$$L\, di/dt = -v_C - Ri$$

which at $t = 0$ becomes

$$di/dt = -4 - 8 = -12$$

Therefore $B = -8$ and the complete solution becomes

$$i = e^{-4t} - 8te^{-4t} = (1 - 8t)e^{-4t} \tag{6.19}$$

A plot of this solution is shown in Figure 6.11.

6.4 A further step

Although it is considered beyond the scope of this book to develop this analysis approach beyond a consideration of the transient response the reader will realize the generality of the approach. If instead of a step excitation a sinusoidal excitation is used the right hand side of the appropriate equation will be a sine term rather than a constant. Solving the equation will produce a complete solution, transient response, and steady state response. Nor is the method restricted to sinusoidal excitation. The mathematics of the solution becomes more difficult as the right hand side terms become less simple.

_____ **Problems** _____

P6.1   Steady state conditions have been attained in the circuit shown in Figure P6.1 when, at $t = 0$, the switch is closed. Calculate:

(a)   a value for the current through the inductance at $t = 0$,
(b)   a value for the current through the inductance as t approaches infinity,
(c)   an expression for the current through the inductance as a function of time.

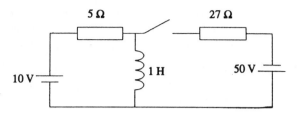

Figure P6.1

P6.2 The circuit shown in Figure P6.2 has reached a steady state with the switch closed when, at $t=0$, the switch is opened. Calculate a value for the voltage, v, across the inductance:

(a)   immediately after the switch is opened,
(b)   at $t=100$ ms.

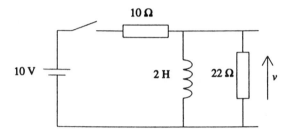

Figure P6.2

P6.3 The circuit shown in Figure P6.2 has reached a steady state with the switch open when, at $t=0$, the switch is closed. The switch is opened again at $t=50$ ms. Calculate a value for the voltage, v, across the inductance at $t=150$ ms.

P6.4 The circuit shown in Figure P6.3 has reached a steady state with the switch closed when, at $t=0$, the switch is opened. Calculate a value for the capacitance voltage after 1 second.

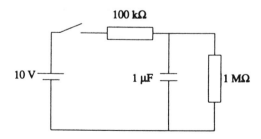

Figure P6.3

P6.5   The switch in the circuit of Figure P6.3 is closed at $t = 0$ and reopened at $t = 100$ ms. Assuming there is no initial charge stored in the capacitance, calculate a value for the voltage across the capacitance at $t = 1$ second.

P6.6   The switch in the circuit of Figure P6.4 is closed at $t = 0$ when the voltage across the 47 nF capacitance is 500 V and that across the 22 nF capacitance is zero. Deduce expressions for the current through and the voltage across the 22 nF capacitance.

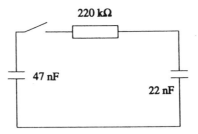

Figure P6.4

P6.7   The circuit shown in Figure P6.5 has no stored energy when, at time $t = 0$, the switch is closed. Deduce an expression for the voltage across the capacitance.

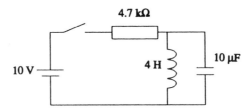

Figure P6.5

P6.8   The circuit shown in Figure P6.6 has no stored energy when, at time $t = 0$, the switch is closed. Deduce an expression for the voltage across the resistance.

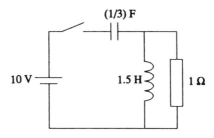

Figure P6.6

P6.9 At time $t = 0$ the current, i, through the inductance of Figure P6.7 is 10 mA and the voltage, v, across the capacitance is -5 V. Deduce an expression for v for $t > 0$.

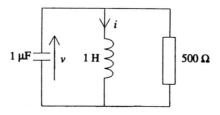

Figure P6.7

P6.10 The capacitance of Figure P6.8 has no stored charge when, at $t = 0$, the switch is moved from position 'a' to position 'b'. At $t = 0.1$ ms the switch is moved to position 'c'. Calculate a value for the voltage across the capacitance at $t = 1$ ms.

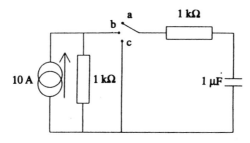

Figure P6.8

CHAPTER SEVEN

Resonance

This chapter introduces the phenomenon of resonance. Simple series and parallel circuits are analyzed and Q factor and bandwidth are both defined and related. A final section looks at some of the finer points arising in lossy resonant circuits.

7.1 Series resonance

The phenomenon of resonance will be demonstrated using numeric examples with a series circuit.

7.1.1 Resonant frequency

Consider the circuit of Figure 7.1.

Calculate the voltages if the circuit current is measured as 50 mA:

$$\omega L = 10^4 \times 100 \times 10^{-3} = 1000$$
$$1/\omega C = 1/(10^4 \times 0.12 \times 10^{-6}) = 833.33$$

Taking the current as the phase reference:

$$\mathbf{V}_R = 50 \times 10^{-3} \times 470 = 23.5 \angle 0° \text{ V}$$
$$\mathbf{V}_L = 50 \times 10^{-3} \times \text{j}1000 = 50 \angle 90°$$
$$\mathbf{V}_C = 50 \times 10^{-3}(-\text{j}833.33) = 41.66 \angle -90°$$

The total circuit impedance is

$$470 + \text{j}1000 - \text{j}833.33 = 470 + \text{j}166.66$$

Therefore the total supply voltage is

$$\mathbf{E} = 50 \times 10^{-3}(470 + \text{j}166.66) = 24.93 \angle 19.53°$$

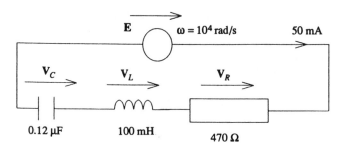

Figure 7.1 The series resonant circuit

It would appear that the voltages across the inductance and the capacitance are greater than the supply voltage. This is sometimes referred to as voltage magnification (see also section 7.1.3). The phasor diagram of Figure 7.2 helps to show how this comes about. It is drawn, only approximately to scale, for the values of Figure 7.1.

As $\mathbf{V}_C$ and $\mathbf{V}_L$ are opposing it is possible for their resultant to be quite small and for the overall resultant voltage to be approximately the same as $\mathbf{V}_R$. In fact it will be seen that if the capacitance value is changed to 0.1 μF then $\mathbf{V}_C$ and $\mathbf{V}_L$ have equal magnitude, the total voltage is equal to $\mathbf{V}_R$, and the circuit current is in phase with the circuit voltage despite the presence of reactive components. This phenomenon is known as *resonance*.

A circuit containing inductance and capacitance is resonant when the circuit current and voltage are in phase.

For the series circuit resonance clearly occurs when $|X_L| = |X_C|$

$$\omega L = 1/\omega C$$
$$\omega = 1/(LC)^{\frac{1}{2}}$$
$$f = 1/2\pi(LC)^{\frac{1}{2}} \tag{7.1}$$

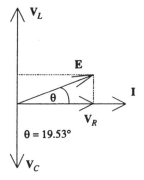

Figure 7.2 Phasor diagram for the series resonant circuit

This is called the resonant frequency of the circuit and is normally denoted f_0. The corresponding angular resonant frequency is denoted ω_0.

7.1.2 Impedance of the series *LCR* circuit

The magnitude of the total impedance of the series circuit is given by

$$|\mathbf{Z}| = [(\omega L - 1/\omega C)^2 + R^2]^{\frac{1}{2}}$$

For given values of L and C plotting $|\mathbf{Z}|$ against frequency for different values of R leads to a set of curves as shown in Figure 7.3. Clearly at resonance the circuit impedance is a minimum.

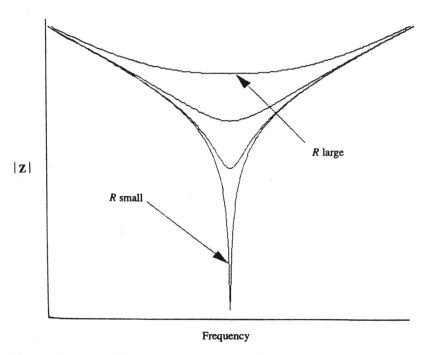

Figure 7.3   Plots of $|\mathbf{Z}|$ against frequency for varying values of R

The sharpest curve corresponds to the smallest value of R. If practical use is to be made of this result it would seem reasonable that the sharpness of the resonance is a property of importance. A quality factor or Q factor for the circuit may be defined *at the resonant frequency* by

$Q = 2\pi \times$ maximum energy stored/total energy lost in one cycle   (7.2)

The higher the value of Q the sharper is the resonance.

This is a completely general definition of Q and more usable expressions may be deduced for particular cases.

7.1.3 Q factor of a series LCR circuit

Let the current through the series circuit be $i = I_{pk} \sin \omega t$.
Energy stored in the inductance, L is $\frac{1}{2}Li^2 = \frac{1}{2}LI_{pk}^2 \sin^2 \omega t$.
Energy stored in the capacitance, C is $\frac{1}{2}Cv_C^2$.
Now

$$v_C = -I_{pk} \cos \omega t / \omega C$$

Therefore energy stored is

$$\frac{1}{2}CI_{pk}^2 \cos^2 \omega t / \omega^2 C^2$$

and hence total energy stored is

$$\frac{1}{2}LI_{pk}^2 \sin^2 \omega t + \frac{1}{2}I_{pk}^2 \cos^2 \omega t / \omega^2 C$$

At resonance $\omega_0^2 = 1/LC$ or $L = 1/\omega_0^2 C$.
Therefore total energy stored is $\frac{1}{2}LI_{pk}^2$.
The average power dissipated is

$$R(I_{pk}/\sqrt{2})^2 = \frac{1}{2}RI_{pk}^2$$

Therefore the energy lost in one cycle at angular frequency ω_0 is

$$\frac{1}{2}RI_{pk}^2 T_0 = \frac{1}{2}RI_{pk}^2 / f_0 = \pi RI_{pk}^2 / \omega_0$$

Hence

$$Q = 2\pi \times (\frac{1}{2}LI_{pk}^2) / (\pi RI_{pk}^2 / \omega_0) = \omega_0 L / R \qquad (7.3)$$

or alternatively

$$Q = 1/\omega_0 CR \qquad (7.4)$$

For the series circuit at resonance the inductive and capacitive reactances are of equal magnitude and opposite sign. Hence the total circuit impedance is equal to the resistance, R. For a supply voltage, E, the circuit current, I, is thus given by

$$I = E/R$$

The voltage across the inductance at resonance is

$$\omega_0 LI = \omega_0 LE/R = QE \qquad (7.5)$$

That is, the voltage across the inductance is Q times the supply voltage. Q is sometimes called the *circuit magnification factor*.
Similarly the voltage across the capacitance is

$$I/\omega_0 C = E/R\omega_0 C = QE$$

7.2 Parallel resonance

Resonance can occur in parallel circuits. The approach to the analysis is the same but the impedance is now found to be a maximum at resonance rather than a minimum. Consider the circuit shown in Figure 7.4.

The total susceptance of the circuit is obtained by adding the inductive and capacitive susceptances as they are in parallel. Thus the

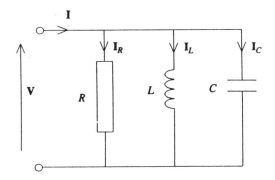

Figure 7.4   A parallel resonant circuit

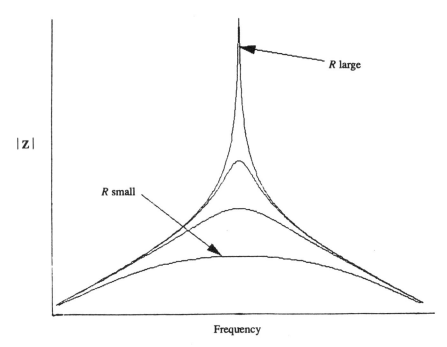

Figure 7.5   Plots of |**Z**| against frequency for a parallel resonant circuit

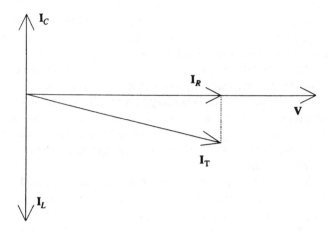

Figure 7.6 A phasor diagram for the parallel resonant circuit

susceptance is

$$1/\omega L - \omega C$$

Invoking equation (5.5) and using $|\mathbf{Z}| = 1/|\mathbf{Y}|$ gives the magnitude of the circuit impedance as

$$|\mathbf{Z}| = 1/[(1/R)^2 + (1/\omega L - \omega C)^2]^{\frac{1}{2}}$$

which when plotted for varying values of R gives the curves of Figure 7.5.

A phasor diagram for this circuit is shown in Figure 7.6.

The total circuit current will be in phase with the circuit voltage when $|\mathbf{I}_C|$ is equal to $|\mathbf{I}_L|$.

Now $|\mathbf{I}_C| = |V|\omega C$ and $|\mathbf{I}_L| = |V|/\omega L$. Therefore resonance occurs when

$$\omega C = 1/\omega L$$

Hence the resonant frequency for this circuit is also

$$\omega_0 = 1/(LC)^{\frac{1}{2}} \quad \text{or} \quad f_0 = 1/2\pi(LC)^{\frac{1}{2}}$$

7.3 Bandwidth of a resonant circuit

If a current of constant amplitude is passed through the circuit of Figure 7.4 the amplitude of the voltage across it will vary with frequency giving a curve with exactly the same shape as Figure 7.5. If the current source contains signals of many frequencies, those in a band of frequencies around the resonant frequency, f_0, will have greater amplitudes than those farther from f_0. This is the process that is used to select between stations in a simple radio receiver. The range of frequencies selected is called the bandwidth of the circuit.

As the response varies smoothly with frequency the boundary between those frequencies selected and those rejected has to be an arbitrary one. In electrical systems this boundary is normally taken as being the frequency at which the power delivered is half the maximum power delivered at the resonant frequency. As power is proportional to the square of the voltage the boundary is also where the voltage has fallen to $1/\sqrt{2}$ times its maximum value.

The bandwidth of the resonant circuit, denoted f_B, is thus the frequency range between the points where the response is $1/\sqrt{2}$ times the response at resonance. The larger the Q value the narrower the bandwidth. It may be shown that the Q value of a resonant circuit is equal to the ratio of the resonant frequency to the bandwidth:

$$Q = f_0/f_B \qquad (7.6)$$

Similar bandwidth considerations can be applied to the series resonant circuit and equation (7.6) still holds. The difference is that the impedance of the series resonant circuit is a minimum at resonance rather than a maximum.

Figure 7.7 illustrates the half-power definition of bandwidth for the series resonant circuit of Figure 7.1. It should be noted that the response curve is not symmetrical about the resonance frequency f_0. The bandwidth $f_B = f_2 - f_1$.

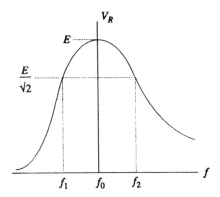

Figure 7.7   Bandwidth of the series *LCR* circuit

_____ **Example 7.1** _____

A capacitor of capacitance $0.1\ \mu F$ is connected in series with an inductor whose inductance is measured as $4\ mH$. Calculate the greatest value of series loss resistance that can be allowed for the inductor to achieve a circuit bandwidth of 500 Hz or less.

$$\omega_0 = 1/(LC)^{\frac{1}{2}} = 1/(4 \times 10^{-3} \times 10^{-7})^{\frac{1}{2}} = 5 \times 10^4\ \text{rad/s}$$
$$\omega_B = 2\pi f_B = 1000\pi$$
$$Q = \omega_0/\omega_B = 50/\pi = \omega_0 L/R$$

Therefore

$$R = \omega_0 L\pi/50 = 50 \times 10^3 \times 4 \times 10^{-3} \times \pi/50 = 12.57\ \Omega$$

7.4 Effects of loss in resonant circuits

In the interests of a simple introduction the preceding sections have analysed the two most obvious circuit configurations to determine their resonant frequencies and the variation with frequency of their imped-ances. The general nature of these results applies to most resonant circuits and they are good approximations quantitatively, provided the Q value of the circuit is sufficiently high. An example of this is the more practical parallel circuit shown in Figure 7.8. This shows a parallel circuit in which the practical inductor is modelled with a series loss resistance rather than a parallel one; this is usually more accurate.

Using 'j' notation to deduce an expression for the resonant frequency it is easier to work with the overall admittance for the parallel circuit:

$$Y_T = j\omega C + 1/(R + j\omega L)$$
$$= j\omega C + (R - j\omega L)/(R^2 + \omega^2 L^2)$$

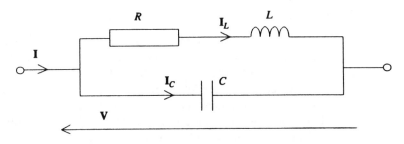

Figure 7.8 An alternative parallel resonant circuit

For resonance the phase angle of $\mathbf{Y}_T$ must be zero, i.e. the j term must be zero:

$$\omega_0 C = \omega_0 L / (R^2 + \omega_0^2 L^2)$$
$$R^2 + \omega_0^2 L^2 = L/C$$
$$\omega_0^2 = 1/LC - R^2/L^2$$
$$f_0 = [1/LC - (R/L)^2]^{\frac{1}{2}}/2\pi \tag{7.7}$$

Note that as R approaches zero the resonant frequency approaches $1/2\pi(LC)^{\frac{1}{2}}$. Also that if $R^2 > L/C$ resonance cannot exist.

Substituting $Q = \omega_0 L/R$, equation (7.7) can be expressed in terms of Q:

$$f_0 = [1/2\pi(LC)^{\frac{1}{2}}][Q^2/(1 + Q^2)]^{\frac{1}{2}} \tag{7.8}$$

Plotting $|Z|$ against frequency for this circuit gives a result very similar to that of Figure 7.5. However, the frequency at which the magnitude of the impedance is a maximum is now given by

$$\omega^2 = (1/LC)[(2 + Q^2)/Q^2]^{\frac{1}{2}} - R^2/L^2 \tag{7.9}$$

which is slightly different from the resonant frequency.

This result is obtained by differentiating an expression for $|Z|$ and is left as an exercise for the reader.

A further example of circuit loss affecting a result is provided by the series circuit of Figure 7.1. If the magnitude of the voltage source is maintained constant but the frequency allowed to vary, the frequency at which the magnitude of the voltage across either reactive component is a maximum may be calculated. It is again left as an exercise in differentiation for the reader to show that the frequency at which the magnitude of the voltage across the inductance is a maximum is given by

$$\omega = (1/LC)^{\frac{1}{2}}[(Q^2 + 0.5)/Q^2]^{\frac{1}{2}} \tag{7.10}$$

_____ **Example 7.2** _____

(a) Calculate the resonant frequency of an ideal parallel resonant circuit consisting of a 25 mH inductance and a 1 μF capacitance.

(b) An inductor of inductance 25 mH and series loss resistance 100 Ω is connected in parallel with a capacitor of capacitance 1 μF. Calculate the resonant frequency of the circuit.

(c) If the number of turns on the inductor of part (b) is increased by a factor of 4 it would be expected that the inductance would increase by a factor of 16 whilst the series loss resistance would increase by a factor of 4 (this assumes that the loss is due solely to the resistance of the wire used). Connect this new inductor in parallel with a capacitor of capacitance $1/16\,\mu F$ so that the *LC* product remains the same. Calculate the resonant frequency of this circuit.

(d) Compare the answers to parts (a), (b), and (c).

(a) The resonant frequency in this case is given by $1/(LC)^{\frac{1}{2}}$. Therefore

$$\omega_0 = 1/(25 \times 10^{-3} \times 1 \times 10^{-6})^{\frac{1}{2}} = 6324.6 \text{ rad/s}$$

or

$$f_0 = 1006.6 \text{ Hz}$$

(b) The resonant frequency is now given by equation (7.6):

$$f_0 = [1/(25 \times 10^{-3} \times 1 \times 10^{-6}) - (100/25 \times 10^{-3})^2]^{\frac{1}{2}}/2\pi$$
$$= 779.7 \text{ Hz}$$

(c) Again using equation (7.6):

$$f_0 = [1/(400 \times 10^{-3} \times 1/16 \times 10^{-6}) - (400/400 \times 10^{-3})^2]^{\frac{1}{2}}/2\pi$$
$$= 993.9 \text{ Hz}$$

(d) The resonant frequency for the circuit of (c) is much nearer to that of the ideal circuit than is that of (b). If the loss of the inductor is mainly due to the resistance of the wire used then it may be concluded that a high L/C ratio for the circuit is a desirable property.

_____ **Problems** _____

P7.1 Calculate values for the inductance and series loss resistance of an inductor to be used in series with a 10 nF capacitance to give a resonant frequency of 10 kHz and a circuit bandwidth of 1 kHz.

P7.2 An a.c. voltage source of magnitude E volts is connected across the circuit of problem P7.1. Show that the voltage across the capacitance at resonance is QE volts.

P7.3 A circuit consists of the parallel combination of an ideal inductance, an ideal capacitance, and an ideal resistance and it resonates at 100 kHz. The circuit impedance is 10 kΩ at resonance and 7.5 kΩ at a frequency of 80 kHz. Calculate the value of the inductance.

P7.4 A 100 mH inductor of series loss resistance 200 Ω is connected in parallel with a capacitor of capacitance 100 nF and the combination driven from a voltage source of impedance 10 kΩ. Calculate the resonant frequency of the parallel circuit. If the frequency of the source is varied calculate the frequency at which the voltage across the capacitor is a maximum.

CHAPTER EIGHT

Mutual inductance and the transformer

This chapter develops a fundamental analysis method for dealing with circuits containing magnetically coupled coils. This enables any such circuit to be analyzed. The ideal transformer is introduced as a specific example of a component involving magnetic coupling. Finally an equivalent circuit which involves only self-inductances is developed for the transformer.

8.1 Mutual inductance

The self-inductance, L, of a coil results from the magnetic field due to a changing current in the coil inducing an e.m.f. in the coil. It was defined in Chapter 3 by the equation $v = L \, di/dt$. Mutual inductance arises where the changing current in a coil induces an e.m.f. in a different coil. At least some of the magnetic flux generated by one coil must link with the second coil. The coils are said to be magnetically coupled. Only two coils are considered for clarity of explanation but the techniques adopted can readily be extended to circuits with more than two coupled coils.

Figure 8.1 shows diagrammatically the two coils to be considered. Their actual physical disposition does not matter as long as the coils are magnetically coupled.

The mutual inductance, M, between the two coils is defined by

$$e_2 = M \, di_1/dt \qquad (8.1)$$

where M is in henries. The mutual inductance is a measure of the 'tightness' of the coupling.

For the given pair of coils if a current i_2 flows in the second coil an e.m.f. is induced in the first coil given by

$$e_1 = M \, di_2/dt \qquad (8.2)$$

As the network is reciprocal M has the same value in each case.

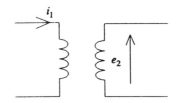

Figure 8.1   Magnetic coupling between two coils

Invoking superposition it may be concluded that in the general case when i_1 and i_2 both exist then voltages e_1 and e_2 are induced in the respective coils.

8.1.1 Dot notation

So far the polarity of the voltages induced has been neglected. If the connections to one coil were reversed, or if the turns of one coil were wound in the opposite direction, then clearly the polarity of the induced voltage would be reversed. In drawing the circuit diagram it is sufficient to have a convention to indicate the polarities resulting from the combined effect of winding direction and method of connection.

A dot is marked at one end of each coil. The interpretation of the dots is that if the current in one coil is flowing *into* the end marked with the dot then the voltage induced in the other coil is *positive* at the end of that coil which is marked with its dot. Should the current be flowing out at the dot of the first coil the voltage will be negative at the dot of the second coil.

When dealing with practical components, measurements must be taken to determine which ends of the windings should be marked as dots when a circuit diagram is drawn. This is determined by the physical orientation of the windings and the direction in which they are wound.

8.2 Analysis involving mutual coupling

The circuit shown in Figure 8.2 is analysed both in the time domain, i.e. with time as the independent variable, and in the frequency domain, i.e. with frequency as the independent variable.

8.2.1 Time domain analysis

The equations introduced in section 8.1 may be applied directly using the time-dependent functions for voltage and current; for circuits with non-sinusoidal excitation the time equations shown below must be used. The

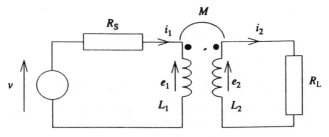

Figure 8.2 Circuit containing a mutual inductance

induced voltages are denoted by the letter e rather than v to emphasize their nature:

$$e_2(t) = M \, di_1/dt$$
$$e_1(t) = -M \, di_2/dt$$

Note that i_1 is labelled flowing into the dot and therefore e_2, as labelled, is positive since the voltage is positive at the dot. The current i_2 is labelled flowing out of the dot resulting in an induced voltage that is negative at the dot of the left hand coil. Thus e_1, as labelled, is negative.

The total voltage appearing across either coil is the algebraic sum of the $L \, di/dt$ term resulting from the self-inductance of the coil and the induced voltage e. For the first loop:

$$v - e_1 = R_S i_1 + L_1 \, di_1/dt$$

or

$$v = R_S i_1 + L_1 di_1/dt - M \, di_2/dt$$

For the second loop:

$$e_2 = R_L i_2 + L_2 \, di_2/dt$$

or

$$0 = R_L i_2 + L_2 \, di_2/dt - M \, di_1/dt$$

Given the excitation, v, and the component values these two equations may be solved for the currents.

8.2.2 Frequency domain analysis

For sinusoidal voltages and currents the 'j' notation may be used and it is easily shown that in this case, still referring to Figure 8.2,

$$\mathbf{E}_2 = j\omega M \mathbf{I}_1 \qquad (8.3)$$

and

$$\mathbf{E}_1 = -j\omega M \mathbf{I}_2 \qquad (8.4)$$

For the first loop:

$$\mathbf{I}_1 = (\mathbf{V} + j\omega M \mathbf{I}_2)/(R_S + j\omega L_1)$$

For the second loop:

$$\mathbf{I}_2 = \mathbf{E}_2/(R_L + j\omega L_2) = j\omega M \mathbf{I}_1/(R_L + j\omega L_2)$$

Again these equations may be solved for the currents.
It is possible to write mesh equations directly:

$$\mathbf{V} = (R_S + j\omega L_1)\mathbf{I}_1 - \qquad j\omega M \mathbf{I}_2$$
$$0 = \qquad -j\omega M \mathbf{I}_1 + (R_L + j\omega L_2)\mathbf{I}_2$$

but care must be taken with the signs.

_____ **Example 8.1** _____

Calculate the input impedance of the circuit shown in Figure 8.3.

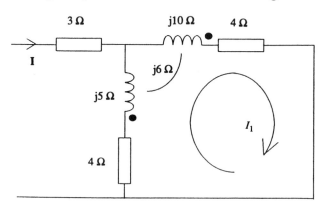

Figure 8.3   The circuit for example 8.1

Let a voltage, V, be applied to the input and apply KVL to each mesh:

$$V = \mathbf{I}(3 + 4 + j5) - \mathbf{I}_1(4 + j5) + j6.\mathbf{I}_1$$
$$= \mathbf{I}(7 + j5) - \mathbf{I}_1(4 - j) \qquad\qquad\qquad (i)$$

and

$$0 = -\mathbf{I}(4 + j5) + \mathbf{I}_1(4 + 4 + j5 + j10) - (\mathbf{I}_1 - \mathbf{I})j6 - \mathbf{I}_1.j6$$

or

$$\mathbf{I}(4 - j) = \mathbf{I}_1(8 + j3) \qquad\qquad\qquad (ii)$$

Substituting (ii) into (i) gives

$$V = (7 + j5)\mathbf{I} - (4 - j)\mathbf{I}(4 - j)/(8 + j3)$$

Now

$$Z_{in} = V/I = 7 + j5 - (16 - 1 - j8)/(8 + j3)$$
$$= (56 - 15 + j40 + j21 - 15 + j8)/(8 + j3)$$
$$= (26 + j69)/(8 + j3)$$
$$= 8.63 \angle 48.8° \ \Omega$$

8.3 Mutually coupled coils in series

The total effective self-inductance of two mutually coupled coils connected in series depends upon the way the coils are connected and the amount of coupling between them.

To calculate the effective self-inductance it is necessary to write an expression for the total voltage across the two coils in the circuit of Figure 8.4. Each coil has a voltage across it due to its self-inductance and a voltage induced in it due to the magnetic coupling to the other coil. Thus there are four terms in all:

$$V = j\omega L_1 I + j\omega L_2 I + j\omega M I + j\omega M I$$

Apply the rule of the dot notation to check that the signs are all positive in this case. Thus

$$V = j\omega(L_1 + L_2 + 2M)I$$

That is, the total effective inductance is $L_1 + L_2 + 2M$ henries.

Now if the connections to one coil are reversed the dot appears at the other end of that coil. It will be seen that the total voltage across the circuit is now given by

$$V = j\omega(L_1 + L_2 - 2M)I$$

Hence in general the self-inductance, L, of two mutually coupled coils in series is given by

$$L = L_1 + L_2 \pm 2M \text{ henries} \tag{8.5}$$

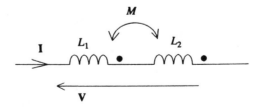

Figure 8.4 Mutually coupled coils in series

This gives a possible way to measure the value of M for two coupled coils. Connect them in series and use a normal measuring bridge to measure the effective self-inductance. Repeat the measurement having reversed the connections to one of the coils. The difference in the two readings is $4M$ henries.

8.4 The transformer

Any two coupled coils constitute a transformer. As the purpose of the transformer is to transmit either information or power it is usual to seek as tight a coupling as possible; thus the coils are wound on a common core.

From energy considerations it can be shown that a maximum value for M exists when the coils are ideally coupled, i.e. no leakage flux exists, and is given by

$$M_{max} = (L_1 L_2)^{1/2} \tag{8.6}$$

where L_1 and L_2 are the self-inductances of the coils.

A *coefficient of coupling*, k, is defined by

$$k = M/M_{max} = M/(L_1 L_2)^{1/2} \tag{8.7}$$

This is an alternative measure of the tightness of the coupling. For ideally coupled coils with no leakage flux k is equal to unity. For coils with no magnetic linkage k is zero and for all practical cases k lies between one and zero. With modern magnetic materials values of k very close to unity can be achieved. However, in applications such as the use of tuned coupled coils in communications, a much lower value of k may be a design requirement.

These results can be used in the analysis of a transformer.

8.4.1 The ideal transformer

Consider the transformer shown in Figure 8.5 and assume no losses of any sort and that k is unity.

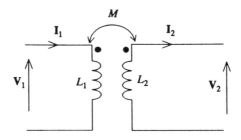

Figure 8.5   The ideal transformer

For the left hand side:

$$V_1 = j\omega L_1 I_1 - j\omega M I_2 \qquad \text{(i)}$$

and for the right hand loop:

$$V_2 = j\omega M I_1 - j\omega L_2 I_2 \qquad \text{(ii)}$$

Multiply (i) by M and (ii) by L_1 and substitute $M = (L_1 L_2)^{1/2}$:

$$V_1 (L_1 L_2)^{1/2} = j\omega L_1 (L_1 L_2)^{1/2} I_1 - j\omega L_1 L_2 I_2$$
$$V_2 L_1 = j\omega L_1 (L_1 L_2)^{1/2} I_1 - j\omega L_2 L_1 I_2$$

Therefore

$$V_1 (L_1 L_2)^{1/2} - V_2 L_1 = 0$$
$$V_2 = (L_2 / L_1)^{1/2} V_1$$

Now $L_1 \propto N_1^2$ and $L_2 \propto N_2^2$ with the same constant of proportionality if both coils are wound on a common core. Therefore

$$V_2 / V_1 = (L_2 / L_1)^{1/2}$$
$$= (N_2^2 / N_1^2)^{1/2}$$
$$= N_2 / N_1 \qquad \text{(8.8)}$$

If a short circuit is connected across the right hand coil of Figure 8.5 the right hand loop then leads to

$$j\omega M I_1 = j\omega L_2 I_2 \quad \text{or} \quad I_1 / I_2 = L_2 / M$$

Substituting $M = (L_1 L_2)^{1/2}$ for the ideal transformer gives

$$I_1 / I_2 = L_2 / (L_1 L_2)^{1/2}$$
$$= (L_2 / L_1)^{1/2}$$
$$= N_2 / N_1 \qquad \text{(8.9)}$$

Thus the voltage ratio across an ideal transformer is equal to the turns ratio and the current ratio is equal to the reciprocal of the turns ratio. In both cases the phase difference between the two voltages, or between the two currents, is zero. If one of the dots in the diagram is moved to the other end of the winding the phase differences are both 180°, i.e. a phase reversal.

8.4.2 Input impedance of an ideal transformer supplying a load Z_L

In Figure 8.6 the input impedance, Z_{in}, of the transformer is given by V_1 / I_1. This may be determined for the ideal transformer by using the

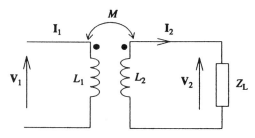

Figure 8.6   A transformer supplying a load Z_L

relationships derived in section 8.4.1:

$$\mathbf{V}_1/\mathbf{I}_1 = (\mathbf{V}_2 N_1/N_2) - (\mathbf{I}_2 N_2/N_1)$$
$$= (\mathbf{V}_2/\mathbf{I}_2) \times (N_1^2/N_2^2)$$
$$= \mathbf{Z}_L N_1^2/N_2^2 \qquad\qquad (8.10)$$

That is, the input impedance is equal to the load impedance multiplied by the square of the turns ratio.

It is left as an exercise for the reader to treat Figure 8.6 as a mutual inductance analysis problem and derive an expression for $\mathbf{V}_1/\mathbf{I}_1$ directly. A hint is to evaluate $\mathbf{Y}_{in}$ rather than $\mathbf{Z}_{in}$ and remember that L_1 and L_2 will be very large for a practical transformer.

Note that the transformer may thus be used to change the magnitude of an impedance, as seen by a source, but it does not change the phase angle. With reference to the maximum power transfer theorem the transformer can be used to obtain a modulus match but not by itself to give a conjugate match.

8.4.3 An equivalent Tee circuit for the transformer

For the circuit of Figure 8.7(a):

$$\mathbf{V}_1 = (R_S + j\omega L_1)\mathbf{I}_1 - j\omega M\mathbf{I}_2$$
$$0 = -j\omega M\mathbf{I}_1 + (R_L + j\omega L_2)\mathbf{I}_2$$

For the circuit of Figure 8.7(b):

$$\mathbf{V}_1 = [R_S + j\omega(L_a + L_c)]\mathbf{I}_1 - j\omega L_c\mathbf{I}_2$$
$$0 = -j\omega L_c\mathbf{I}_1 + [R_L + j\omega(L_c + L_b)]\mathbf{I}_2$$

These two sets of equations are identical if

$$L_1 = L_a + L_c \quad M = L_c \quad L_2 = L_c + L_b$$

which gives

$$L_a = L_1 - M \quad L_b = L_2 - M \quad L_c = M$$

Therefore the equivalent circuit becomes as shown in Figure 8.8.

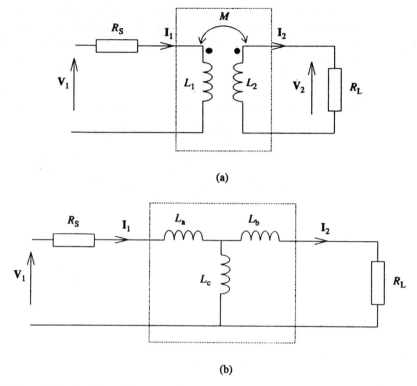

(a)

(b)

Figure 8.7 Equivalent Tee circuit for a transformer

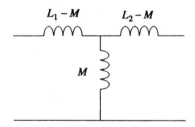

Figure 8.8 Inductance values of the equivalent Tee network for a transformer

If one of the transformer windings is reversed, i.e. one of the dots appears at the other end of the winding, then it is easily shown that the values for the equivalent circuit inductances are given by

$$L_a = L_1 + M \qquad L_b = L_2 + M \qquad L_c = -M$$

Note that this equivalent circuit will result in a correct analysis even though it has a negative value for inductance which cannot be directly implemented. At a single frequency this could be implemented with a capacitance but this is seen to be only an analytic model as the capacitance required is a frequency-dependent capacitance given by $C = 1/\omega^2 M$.

_____ **Problems** _____

P8.1   Calculate the current **I** and the voltage **V** in Figure P8.1.

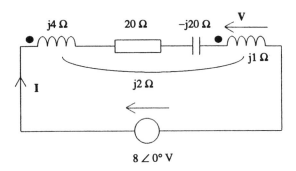

Figure P8.1

P8.2   Calculate the voltage **V** in Figure P8.2.

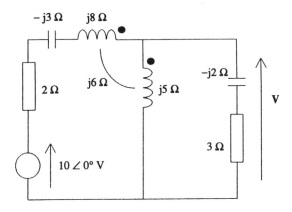

Figure P8.2

P8.3   Calculate the current **I** in Figure P8.3.

P8.4   For the circuit of Figure P8.4, calculate the mutual inductive reactance, ωM, which would result in a power dissipation of 45.2 W in the 5 Ω resistor.

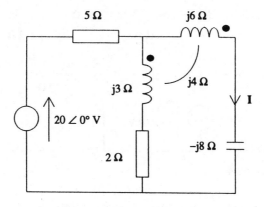

Figure P8.3

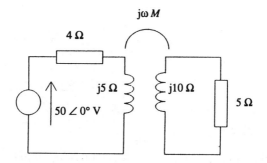

Figure P8.4

P8.5 Calculate the current drawn from the source in Figure P8.5 if $k = 0.5$.

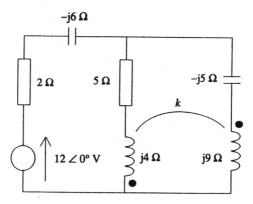

Figure P8.5

P8.6 Calculate the current in each resistor of Figure P8.6.

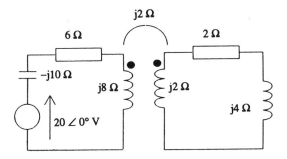

Figure P8.6

P8.7   Calculate the impedance between points 'a' and 'b' of the circuit of Figure P8.7.

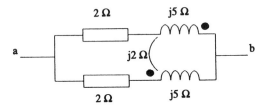

Figure P8.7

P8.8   For the circuit of Figure P8.8 calculate the voltage across the capacitor and the value of the total impedance connected across the source.

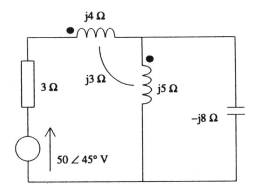

Figure P8.8

P8.9   Calculate a Thévenin equivalent circuit for the circuit of Figure P8.9 between terminals 'a' and 'b'.

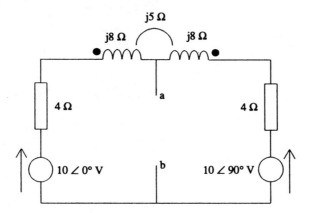

Figure P8.9

Fields

In free space in the vicinity of charges, currents, and magnets, forces are experienced on other charges, currents, and magnets. We have what is commonly known as *action at a distance*. This is not new; gravity has always demonstrated this type of action. There is nothing tangible in this space and so we postulate the existence of electric and magnetic fields to account for the forces.

The following chapters describe the properties and effects of these fields. It is common practice to sketch the fields by field lines in space, and to interpret the number of lines per unit area as a measure of the field strength. The electric and magnetic fields may exist separately or coexist in the same spatial volume, and may be constant or variable in time. When the fields coexist and are dependent on each other they are called *electromagnetic fields*.

It is one of the great achievements of theoretical physics, due to James Clerk Maxwell, that four fundamental equations are able to explain all the non-relativistic phenomena associated with electromagnetic fields. The following chapters aim to develop an understanding of electric and magnetic fields and lead to a simple presentation of the four equations in Chapter 12.

CHAPTER NINE

Electric fields

Electrostatics is the study of the electric field due to the presence of stationary charges. The charges may be negative or positive. This is called the *polarity* of the charge. In general bodies are neutral, i.e. they do not have a residual charge. The presence of charge is due to the excess of electrons giving a negative charge, or the deficiency of electrons giving a positive charge. The electronic charge is therefore fundamental and has a value of -1.6×10^{-19} coulombs.

Materials for electromagnetic purposes divide into the two main groups, *conductors* and *insulators*. In conductors electrons flow easily and the material is said to have a high conductivity. Copper is an example of a very good conductor and has a conductivity of 5.8×10^7 siemens/m. In insulators, also called *dielectrics*, the electrons cannot flow easily and the material is said to have a low conductivity. Quartz is an example of a good insulator and has a conductivity of 10^{-17} siemens/m. This conductivity range is extremely large.

Experimentally it is found that forces exist between charges. An important feature of *like* (same polarity) charges are that they repel each other; *unlike* (opposite polarity) charges attract each other.

In a *closed* system, where no charge enters or leaves the system, charge is *conserved*, i.e. no charge can be destroyed or created. Thus if a system is neutral, although regions of positive and negative charge can be created the total charge must remain zero. This is an important concept and is known as the *conservation of charge*.

9.1 Coulomb's law

Experimentally it is found that (see Figure 9.1):

The force between two point charges in free space is proportional to the product of the two charges, inversely proportional to the square of the distance apart, and acts along the line between the charges.

Figure 9.1   Two point charges distance *r* apart

This may be written

$$F = k q_1 q_2 / r^2 \tag{9.1}$$

where k is a constant. The constant k is chosen to have a value $1/4\pi\varepsilon_0$ to ensure that all units agree with the SI system. Thus force will be in *newtons* when charge is in *coulombs* and distance in *metres*. The value 4π ensures that other derived expressions are simpler and also indicates spherical symmetry. In addition if vectors are used to specify the correct direction of the force the equation becomes

$$\mathbf{F} = q_1 q_2 \hat{\mathbf{r}} / 4\pi\varepsilon_0 r^2 \tag{9.2}$$

The force on q_1 and q_2 will be attractive if they have opposite polarity, and repulsive if they have the same polarity. Equation (9.2) is Coulomb's law.

The value of the constant ε_0 is called the *permittivity of free space* and has the value 8.85×10^{-12} farads/m. If the charges are in a dielectric other than free space then the value will change to a value ε for the new material. This may be expressed in terms of a relative permittivity ε_r where

$$\varepsilon = \varepsilon_r \varepsilon_0 \tag{9.3}$$

Typical values of ε_r are: air, 1.0006; glass, 5; mica, 4.

_____ **Example 9.1** _____

Find the force acting on a $1\ \mu C$ charge at P due to the two charges at A and B as shown in Figure 9.2.

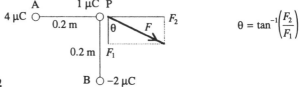

Figure 9.2

The forces are found using Coulomb's law, and the results for A and B added vectorially.

The force due to A on P is

$$F_2 = (4.10^{-6})(1.10^{-6})/4\pi\varepsilon_0(0.2)^2 = 0.9 \text{ N}$$

This is a repulsive force and therefore the force is to the right.
The force due to B on P is

$$F_1 = (2.10^{-6})(1.10^{-6})/4\pi\varepsilon_0(0.2)^2 = 0.45 \text{ N}$$

This is an attractive force and therefore acts downwards.

The resultant is found by adding vectorially, i.e. completing the parallelogram and calculating F:

$$F = (F_1^2 + F_2^2)^{\frac{1}{2}} = 1.006 \text{ N}$$

and angle $\theta = \tan^{-1}2 = 63.4°$

9.2 Electric field strength

To explain the ability of electric forces to act at a distance from point charges we postulate the existence of an electric field permeating space around the charges. By measuring the strength and direction of forces we can obtain 'pictures' of the field. Figure 9.3 shows pictures where the field is represented by lines.

The spacing of the lines shows the strength of the field. Note how the field lines commence on a positive charge and terminate on a negative charge or at infinity (Figure 9.3(c)). In Figure 9.3(c) if the charge were negative the field lines would point inwards. For an isolated charge the field has *spherical symmetry*.

Now suppose a small test charge q is placed at A in a field E (Figure 9.4). Then the field strength at A is defined as *the force per unit charge at A*, and will have the same direction as the force if the charge q is positive.

Thus $E = F/q$ and rearranging the equation and writing in vector form

$$\mathbf{F} = q\mathbf{E} \tag{9.4}$$

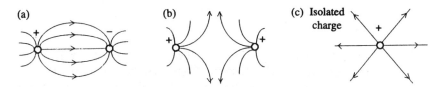

Figure 9.3   Field lines for some charge configurations

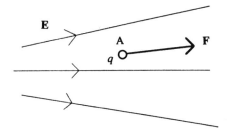

Figure 9.4   Force on a positive charge *q* in a field **E**

When *F* is in newtons and *q* in coulombs then the units for *E* will be volts/m.

9.2.1 Field due to an isolated point charge

Consider a point charge *q* as in Figure 9.3(c). A test charge of unit magnitude placed at distance *r* from *q* will experience a force *F* given by Coulomb's law of $q/4\pi\varepsilon_0 r^2$. From equation (9.4) the force must have the same magnitude as the field since the test charge has unit magnitude. Therefore the expression for the field in vector form is

$$\mathbf{E} = q\hat{\mathbf{r}}/4\pi\varepsilon_0 r^2 \qquad (9.5)$$

The field will be radially outward if the charge *q* is positive.

_____ **Example 9.2** _____

Find an expression for the field strength at P due to an electric dipole as shown in Figure 9.5.

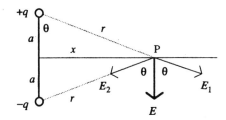

Figure 9.5

If a unit charge is placed at P, then the forces at P will have the same directions as the field strengths. There will be a repulsive force due to +*q* and an attractive force due to −*q*. The fields add vectorially to give a resultant field acting vertically down.

The field at P due to each charge is

$$E_1 = E_2 = q/4\pi\varepsilon_0 r^2$$

The total field at P will be given by

$$\mathbf{E} = \mathbf{E}_1 + \mathbf{E}_2$$

From Figure 9.5 the magnitude is

$$E = 2E_1 \cos\theta$$

and since $\cos\theta = a/r$

$$E = qa/2\pi\varepsilon_0 r^3$$

E acts vertically down. If the electric dipole is very small compared with the distance x then $a \lhd x$. Then the result for E is approximately given by $qa/2\pi\varepsilon_0 x^3$.

_____ **Example 9.3** _____

Find the field at P, distance a from a line charge of length b, if the charge density on the line is λ coulombs per unit length. P is along the length of the line.

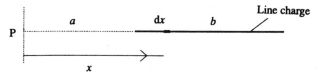

Figure 9.6

Consider P as shown in Figure 9.6, and a small element dx on the line distance x from P. P is chosen on the left so that the distance x from P to the element dx is in the positive direction. The charge on the element dx will be $\lambda\, dx$. The field at P due to dx is, by Coulomb's law,

$$dE = \lambda\, dx/4\pi\varepsilon_0 x^2$$

The field due to the line is

$$E = (\lambda/4\pi\varepsilon_0) \int_a^{a+b} dx/x^2$$
$$= (\lambda/4\pi\varepsilon_0)[-1/x]_a^{a+b}$$
$$= \lambda b/4\pi\varepsilon_0 a(a+b)$$

If $b \gg a$ then this expression becomes approximately $\lambda/4\pi\varepsilon_0 a$.

9.2.2 Fields due to continuous charge distributions

It is possible to represent complex charge distributions by a large number of point charges. The total field can then be found by integrating over all the point charges. The integration can be difficult, but there are several symmetrical situations where it is easy. Two examples of the 'point charge' method of calculating electric field strength will be given.

It is important to understand that the *differential element*, e.g. d*l* where *l* represents length, associated with the integration enters the integration *naturally* due to the physics of the problem.

─────── **Example 9.4** ────────────────────────────────

Consider a line charge of magnitude λ C/m that is bent into a circle of radius a (Figure 9.7). What is the field at point P distance x from the circular line charge?

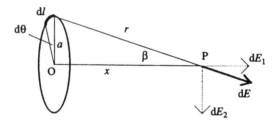

Figure 9.7

Consider a small element d*l* of the line; this will be the 'point charge' and will have a charge $\lambda \, dl$. The element $dl = a \, d\theta$ where $d\theta$ is the angle subtended at O by d*l*.

By Coulomb's law this element will produce a field dE at P of value

$$dE = \lambda a \, d\theta / 4\pi\varepsilon_0 r^2$$

dE may be resolved into dE_1 and dE_2. The field dE_2 is always cancelled by an element diametrically opposite and therefore only the horizontal component dE_1 is left. The resultant field will be due to contributions from all the elements around the line charge, and this may be found by integration.

The field along x is

$$E_1 = \int dE \cos \beta$$

$$= \int_0^{2\pi} (\lambda a x \, d\theta / 4\pi\varepsilon_0 r^3) \quad \text{since } \cos \beta = x/r$$

$$= (\lambda a x / 4\pi\varepsilon_0 r^3) \int_0^{2\pi} d\theta \quad \text{since } a, x, \text{ and } r \text{ are constant}$$

$$= \lambda a x / 2\varepsilon_0 r^3$$

Note that dθ entered the problem naturally via the small element d*l*.

An alternative simpler method for finding the field strength in symmetrical situations using Gauss's law will be dealt with in section 9.6.

9.2.3 Field due to an infinitely long straight line charge

This is a useful result since it may be used to find the field due to other charge distributions. The result can be found using the method shown in the examples of section 9.2.1 but the integration is rather messy. (The reader should try it if his or her integration is good!)

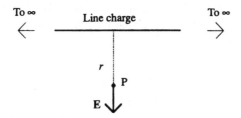

Figure 9.8 Field at P due to an infinitely long straight line charge

As mentioned above there is a simpler method using Gauss's law, and this problem will be solved in section 9.6. For the present the result will be stated only. The expression for this field, at a point P distance r from the line (Figure 9.8), when the charge on the line is λ coulombs per unit length, is given by

$$\mathbf{E} = \lambda \hat{\mathbf{r}}/2\pi\varepsilon_0 r \tag{9.6}$$

and will point radially outwards if the charge is positive.

Note that when there is *cylindrical* symmetry the value 2π is present in the denominator.

9.3 Electrical potential

Electrical potential is normally derived from electrical potential difference. Their definitions are based on work done and therefore both are *scalar quantities* and usually use the symbol V. Electrical potential is very important in field studies and as we shall see is intimately related to electric field.

9.3.1 Potential difference

The potential difference between two points a and b is defined as

$$V_b - V_a = W_{ab}/q \tag{9.7}$$

where V_b and V_a are the potentials at b and a and W_{ab} *is the work done by an external agent in moving a charge q from a to b.* The units of V are volts. It may be shown that the potential difference thus defined is *independent of the path* taken between *a* and *b*. Note that when W_{ab} and *q* are positive then $V_b > V_a$.

9.3.2 Potential at a point

Suppose point *a* is removed to infinity and *a* is defined to have zero potential making $V_a = 0$; then we have

$$V_b = W_{ab}/q \tag{9.8}$$

and W_{ab} *is the work done by an external agent in moving a charge q from infinity to the point b.* Infinity is chosen for zero potential since at such a large distance the effects of any charge distributions can be assumed to be negligible.

9.4 Potential and field strength

Consider a field **E** pointing from left to right as shown in Figure 9.9. In such a situation $V_b > V_a$ since the electric field points from high to low values of potential. A charge *q* is placed at point P. Then a force $F = qE$ will be exerted to the right on the charge by the field. An external agent will have to do positive work against the field **E** in order to push the charge from *a* to *b*. The work done to push *q* over a small distance d*l* is given by the force required to overcome the force due to the field **E** *resolved* along the direction **dl**. The scalar product can be used to provide a resolving effect without the need for a detailed diagram showing angles. The work done by the external agent is

$$W_{dl} = q\mathbf{E} \cdot \mathbf{dl}$$

Therefore

$$W_{ab} = -\int_a^b q\mathbf{E} \cdot \mathbf{dl}$$

and

$$V_b - V_a = -\int_a^b \mathbf{E} \cdot \mathbf{dl} \tag{9.9}$$

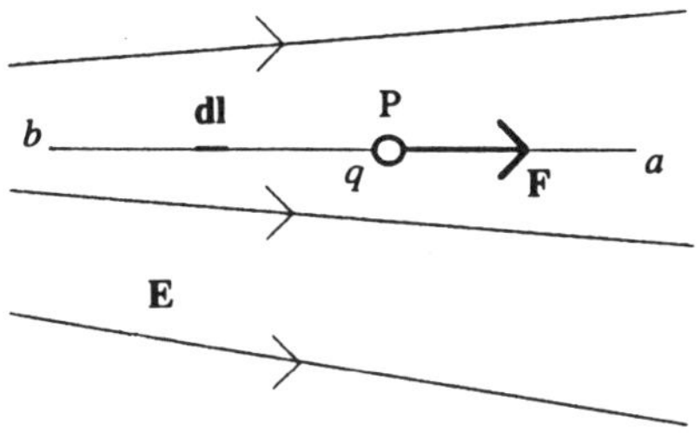

Figure 9.9 Work is done to move a positive charge from *a* to *b* against the field **E**

The minus sign needs further explanation. $V_b - V_a$ is positive since the field **E** points from left to right. However **dl** points right to left for q is being pushed from a to b. Therefore a minus sign is required to make the signs correct on both sides of equation (9.9).

The right hand side of equation (9.9) is known as a *line integral*. This equation emphasizes the fact that if the unit of potential is the volt then electric field will have the unit volts/m. Considering a path ab which is very small and of length dl, and letting $V_b - V_a = dV$, the equation becomes $dV = -E'dl$, where E' is the field component in the dl direction. This may be written as $E' = -dV/dl$ and is the *negative potential gradient* in the dl direction.

The *maximum* value of the negative potential gradient at a point in the electric field gives the field **E** in magnitude and direction at that point.

9.4.1 Potential of a point charge

The potential at P distance r from an isolated point charge q is a useful result since distributions of charge can be made up from point charges (Figure 9.10).

From the basic definition we know that the potential is the work done to bring a unit test charge from infinity to point P. Consider the situation when the unit test charge is at N, an arbitrary distance x from the charge q. At N the field due to q is $q/4\pi\varepsilon_0 x^2$ and the force on the unit test charge has the same value since force = field × charge. Work done is force × distance; therefore the work done to push the test charge a small distance dx towards q against this force is given by the expression

$$\text{Work done} = -q\,dx/4\pi\varepsilon_0 x^2$$

The minus sign is needed since the test charge is being pushed in the negative direction of x. The potential is the work done per unit charge to move from infinity to P and our test charge has unity magnitude. Therefore the potential at P is

$$-\int_\infty^r q\,dx/4\pi\varepsilon_0 x^2 = q/4\pi\varepsilon_0 r \tag{9.10}$$

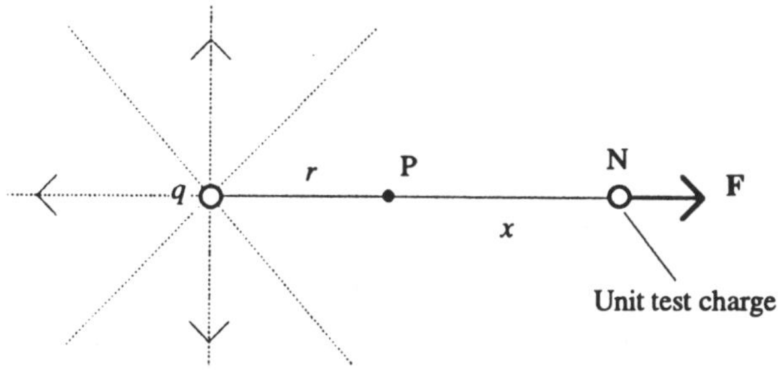

Figure 9.10 A unit test charge is moved from infinity to the point P

This is a scalar quantity. For several point charges the potential at a point is found by adding the individual contributions of each charge. Each contribution is calculated as though the other charges were not present. The electric field at distance r from a point charge q may be found, from the potential, by taking the negative potential gradient of $q/4\pi\varepsilon_0 r$ along the r direction. This gives the result $q/4\pi\varepsilon_0 r^2$ as expected. Two examples based on equation (9.10) follow.

Example 9.5

What is the potential at P due to the three point charges arranged as in Figure 9.11?

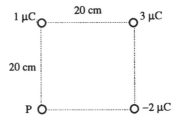

Figure 9.11

Each point charge has a potential at P given by equation (9.10). It is only necessary to superpose these potentials to find the resultant value at P. They are scalars so direction is not involved and the addition is simple:

$$V = (1/4\pi\varepsilon_0)\{(10^{-6}/0.2) + [3.10^{-6}/(0.2)(1.414)] - (2.10^{-6}/0.2)\}$$
$$= (5.61)10^{-6}/4(3.142)(8.85)10^{-12}$$
$$= (50.4)10^3 \text{ V}$$

Potentials tend to be high in electrostatics even for small values of charge owing to the presence of ε_0 in the denominator.

Example 9.6

Find the potential at point P in Figure 9.12 along the axis and distance 10 cm from a charged disk of radius 10 cm. The disk has a uniform surface charge of value 10^{-6} C.

The charge density on the disk is

$$\sigma = 10^{-6}/\pi(0.1)^2 = (3.18)10^{-5} \text{ C/m}^2$$

Consider an annulus on the disk radius a and width da. This annulus will have a charge $\sigma 2\pi a$ da. All points on this annulus are the same distance r from P. Therefore the potential at P is obtained by adding contributions from all the small annuli from the centre to the radius 0.1 m.

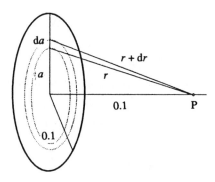

Figure 9.12

The potential due to the annulus is

$$dV = \sigma 2\pi a \, da / 4\pi\varepsilon_0 r$$

The potential due to the disk is

$$V = \int_0^{0.1} \sigma 2\pi a \, da / 4\pi\varepsilon_0 r$$

This integral is easy if the variable r is used instead of a. Note that $a^2 = (r^2 - 0.01)$ and $2a \, da = 2r \, dr$.

Substituting these values into the above expression, the potential due to the disk is

$$V = (\sigma/2\varepsilon_0) \int_{0.1}^{(0.1)\sqrt{2}} dr$$

$$= (74.4)10^3 \text{ V}$$

9.4.2 Equipotentials and field lines

An equipotential is a line joining all points in a field that have the same potential. From a practical point of view the equipotentials are useful because they are usually easily determined and the field lines **E** can be estimated from them. Equation (9.9) gives the relationship between potential difference and field as

$$V_b - V_a = -\int_a^b \mathbf{E} \cdot \mathbf{dl}$$

If we move perpendicular to the field lines then **dl** will be at right angles to **E**. For this condition the scalar product **E·dl** is always zero, indicating *zero* potential difference. In other words, *equipotential lines are at right angles to field lines.*

Equipotential lines close together mean a high potential gradient, and this implies a large field with close field lines. It is therefore possible to

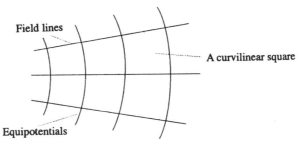

Figure 9.13 Field lines and equipotentials are at right angles enclosing curvilinear squares

sketch field lines from equipotentials using two properties:

1. The two sets of lines must be at right angles.
2. The two sets of lines should form '*curvilinear squares*' to maintain high density field lines when equipotentials are of high density and vice versa. A curvilinear square is a rough square bounded by curved and/or straight lines (Figure 9.13).

The equipotentials may be obtained by voltmeter methods and a sketch for the field can then be made. *Field plotting* methods like this, utilizing graphite resistive sheets, can be used to find field patterns for complicated conductor shapes.

9.5 Electric flux and flux density

Flux is an important concept and may be applied to any vector quantity. In section 9.5.1 the flux associated with electric field strength **E** is found. It is called the flux of **E**, and must not be confused with electric flux which is reserved for another electric quantity! (see section 9.5.2). Flux is a *scalar* quantity since total amounts irrespective of direction are usually considered.

9.5.1 The flux of **E**

Consider lines of uniform **E** crossing a surface S in Figure 9.14, such that the angle between **E** and the normal to the surface is θ.

Then the flux across the surface can be defined as the normal component of **E** multiplied by the surface area S or flux $= ES \cos \theta$. Note that when **E** is parallel to the surface and $\theta = 90°$ then the flux is zero. This is clearly a sensible result.

If the electric field is not uniform, then E can vary in strength over the area. Then we must treat a small area ds and integrate over the surface

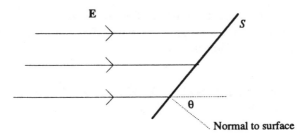

Figure 9.14 The flux of **E** crossing a surface *S*

area *S* to get the total flux. By definition a vector **ds** has a small area value d*s*, and is normal to the surface. The scalar product can be used to resolve along the normal, and therefore we have

$$\text{The flux of } \mathbf{E} = \int_S \mathbf{E} \cdot \mathbf{ds} \tag{9.11}$$

If the surface *S* is closed then

$$\text{The flux of } \mathbf{E} = \oint_S \mathbf{E} \cdot \mathbf{ds} \tag{9.12}$$

_____ **Example 9.7** _____

A cube is immersed in a uniform field with the field perpendicular to faces ABCD and EFGH in Figure 9.15. If the faces have area *S* what is the total flux flowing out of the cube?

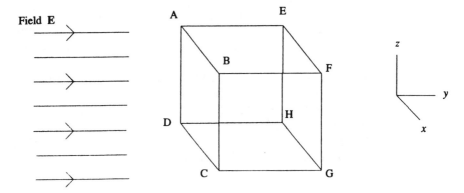

Figure 9.15

All six sides must be dealt with independently for they have different 'directions'.

Since the field is uniform the expression from equation (9.11) for one side degenerates to $S(\mathbf{E} \cdot \mathbf{n})$ where S is the area of a side and $\mathbf{n}$ is the normal out of the surface.

Total flux out
$$= \text{flux from (ABCD + EFGH + AEFB + CDHG + AEHD + BFCG)}$$
$$= S(-\mathbf{E} \cdot \mathbf{a}_y + \mathbf{E} \cdot \mathbf{a}_y + \mathbf{E} \cdot \mathbf{a}_z - \mathbf{E} \cdot \mathbf{a}_z - \mathbf{E} \cdot \mathbf{a}_x + \mathbf{E} \cdot \mathbf{a}_x)$$
$$= 0$$

The last four terms are zero since $\mathbf{E}$ is perpendicular to z and x, and the first two terms cancel. The zero result should not be surprising, since flux is not created or destroyed within the cube. How would the problem be dealt with if the cube was oriented at an arbitrary angle to the field?

9.5.2 Electric flux density

Although $\mathbf{E}$ is sufficient to describe the electric field, it is common practice to define another vector quantity $\mathbf{D}$ which is related to $\mathbf{E}$ by the expression

$$\mathbf{D} = \varepsilon\mathbf{E} = \varepsilon_0\varepsilon_r\mathbf{E} \tag{9.13}$$

$\mathbf{D}$ is called the *electric flux density* and has the units of C/m^2. $\mathbf{D}$ always has the same direction as $\mathbf{E}$ but changes in magnitude according to the dielectric material.

$\mathbf{E}$ depends on all charges present, which includes free charge and bound charge. Bound charge is the charge that is tied to the atoms of the dielectric. $\mathbf{D}$ depends on the *free charge only*. This means that $\mathbf{D}$ can be easier to use than $\mathbf{E}$.

The simplicity of equation (9.13) hides the true nature of $\mathbf{D}$, which is associated with the polarization of the dielectric. More information on $\mathbf{D}$ may be found in Appendix 5. An alternative name for $\mathbf{D}$ used in older textbooks is the *displacement*.

The total electric flux uses the symbol Ψ, and is found by integrating $\mathbf{D}$ over area. The units for Ψ are coulombs. Therefore

$$\Psi = \int_S \mathbf{D} \cdot \mathbf{ds} \tag{9.14}$$

All equations containing $\mathbf{E}$ may be rewritten using $\mathbf{D}$ by utilizing equation (9.13). In this text, however, $\mathbf{E}$ will be used almost exclusively.

9.6 Gauss's law

Gauss's law for electrostatics may be stated as:

The flux of $\mathbf{E}$ across a closed surface S enclosing a net charge q is equal to q/ε.

This can be written as

$$\oint_S \mathbf{E} \cdot \mathbf{ds} = q/\varepsilon \qquad (9.15)$$

In free space (or air) ε is replaced by ε_0.

This is a very important relationship and may be used to calculate $\mathbf{E}$ when the charge distribution is known. The surface S is chosen to make analysis easy by utilizing symmetry. The value of E obtained will be the value at the Gaussian surface. Examples to demonstrate the law follow.

_____ **Example 9.8** _____

Using Gauss's law determine the field at distance r from an isolated point charge q in free space.

This situation has spherical symmetry and therefore a sensible shape for the Gaussian surface is a sphere (Figure 9.16). The field is required at distance r from the charge and therefore the Gaussian surface must be placed at radius r.

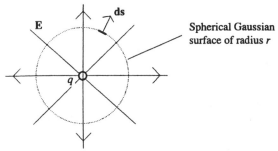

Figure 9.16

By Gauss's law

$$\oint_S \mathbf{E} \cdot \mathbf{ds} = q/\varepsilon_0$$

On the sphere $\mathbf{E}$ is parallel to $\mathbf{ds}$ and therefore $\mathbf{E} \cdot \mathbf{ds}$ becomes $E\,ds$.

Symmetry suggests that E is constant over the sphere and therefore may be taken outside the integral sign. Therefore the above expression becomes

$$E \oint_S ds = q/\varepsilon_0$$

Integrating

$$E4\pi r^2 = q/\varepsilon_0$$

giving

$$E = q/4\pi\varepsilon_0 r^2$$

This result agrees with the value quoted previously. The direction is radially outwards if q is positive and this could be indicated by normal vector notation.

——— **Example 9.9** ——————————————————————

Find the field strength at distance r from an infinite line charge of density λ C/m.

The problem has cylindrical symmetry. The field is required at radius r. A cylindrical Gaussian surface of radius r is therefore chosen (Figure 9.17).

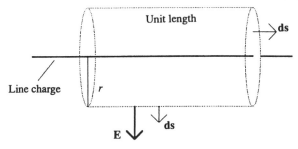

Figure 9.17

Consider a unit length of line; then the charge on this length will be λ. Symmetry suggests the field must be radial. On the ends of the cylinder the outwards normal **ds** is perpendicular to the field. Therefore $\mathbf{E} \cdot \mathbf{ds} = 0$ and there is no contribution from the ends. On the curved portion **ds** is parallel to **E** and therefore $\mathbf{E} \cdot \mathbf{ds} = E\,ds$.

It is not necessary to choose a shape for the elemental ds since the sum of all the ds must be the total area of the curved portion of the unit length cylinder. Therefore

$$\int_s \mathbf{E} \cdot \mathbf{ds} = \int_s E\,ds = E \int_s ds = E2\pi r$$

and

$$E2\pi r = \lambda/\varepsilon_0$$

giving

$$E = \lambda/2\pi\varepsilon_0 r$$

This agrees with the value quoted in section 9.2.3. If required E may be specified as a vector in the usual way. Note how easy the problem is when symmetry exists and Gauss's law can be applied.

If the charge density is 0.2 μC/m and the field is desired at a distance 20 cm from the line, the value for the field is approximately 18 000 V/m.

9.6.1 Gauss's law and continuous charge distributions

The charge in a region may exist as a cloud of charge. Then we specify the value as a volumetric charge density with units of C/m^3. If the charge density is ρ at a small volume dv, then the total charge in a volume V can be described as $\int_V \rho \, dv$ and Gauss's law becomes

$$\oint_S \mathbf{E} \cdot \mathbf{ds} = (1/\varepsilon) \int_V \rho \, dv \tag{9.16}$$

where S is the surface surrounding the volume V.

A value for the right hand side of this equation would be difficult to calculate unless the charge was distributed simply and symmetrically.

9.7 Capacitors and dielectrics

Two conductors carrying equal and opposite charges with a dielectric between form a capacitor (Figure 9.18).

Electric field lines leave the positive charge and terminate on the negative charge. The potential difference between the charges can be found by taking *any* path between the conductors and finding the voltage drop across it. This can be done using the standard formula for potential difference which is repeated here as $V_b - V_a = -\int_c \mathbf{E} \cdot \mathbf{dl}$. For simplicity let this potential difference be called V.

Then the capacitance of the system of the two charges is defined as

$$C = q/V \tag{9.17}$$

The units for C are farads when q is in coulombs and V in volts. The

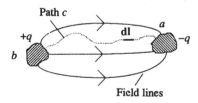

Figure 9.18 Two conductors forming a capacitor. The path may be any shape

symbol for farads is F. The farad is a rather large unit, and for most practical applications values in μF and even smaller units are used.

9.7.1 The parallel plate capacitor

For the capacitor shown in Figure 9.19 let $A \geqslant d$, in which case we can ignore end effects and assume a uniform field between the plates. Let the dielectric constant of the material between the plates be ε.

C can be obtained by finding the relationship between E and V and a value for E in terms of q. Applying equation (9.9) where $V_b = 0$ and $V_a = V$, and the path is a straight line between the plates,

$$0 - V = -\int_0^d \mathbf{E} \cdot \mathbf{dl}$$

$\mathbf{E}$ and $\mathbf{dl}$ are parallel, therefore $\mathbf{E} \cdot \mathbf{dl} = E \, dl$, and since E is constant it may be taken outside the integral sign. Therefore

$$V = E \int_0^d dl$$

$$= Ed \tag{9.18}$$

This is a simple well-known result for a uniform field.

Now consider a Gaussian surface around the top plate. The surface should be a closed 'box', one that surrounds the plate so that the charge contained is $+q$. The field is uniform and the normal from the surface is parallel to $\mathbf{E}$. Therefore

$$\int \mathbf{E} \cdot \mathbf{ds} = E \int ds = EA$$

Note that there is no contribution from the sides of the box since over these regions $\mathbf{E}$ is perpendicular to $\mathbf{ds}$. There is also no contribution from the top of the box since $\mathbf{E} = 0$ there. By Gauss's law

$$EA = q/\varepsilon$$

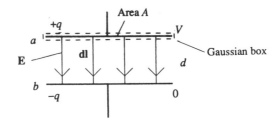

Figure 9.19 A parallel plate capacitor

Substituting equation (9.18) and the above value for E into $C = q/V$ gives an expression

$$C = \varepsilon A/d \qquad (9.19)$$

For $A = 10$ m^2, $d = 0.1$ mm, and $\varepsilon_r = 5$, the value of C is 4.43 μF.

———— **Example 9.10** ————————————————————————————

Find the capacitance of the 'sandwich' capacitor shown in Figure 9.20. The dielectric slab has a relative permittivity of 4. The dimensions are $A = 1$ m^2, $a = 0.2$ mm, $b = 0.1$ mm.

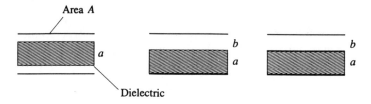

Figure 9.20

The diagram on the left may be replaced by the centre one, for the position of the slab is not important. The presence of a metal plate on top of the slab as shown in the right diagram also does not affect the capacitance. Thus the original capacitor can be replaced by two capacitors, one with air dielectric and one with the slab dielectric. These two capacitors are in series. The capacitances can be calculated and then added according to the expression $C = C_1 C_2/(C_1 + C_2)$:

$$C_1 = \varepsilon_r \varepsilon_0 A/a = 4(8.85.10^{-12})/2.10^{-4} = 177 \text{ nF}$$
$$C_2 = \varepsilon_0 A/b = 8.85.10^{-12}/10^{-4} = 88.5 \text{ nF}$$

Therefore

$$C = 59 \text{ nF}$$

9.7.2 Capacitance of a coaxial cable

Coaxial cable is used extensively in telecommunications. The capacitance per unit length is an important attribute.

In Figure 9.21 the outer and inner conductors are made from copper and have radii a and b. A dielectric with permittivity ε is placed between the conductors. The charge per unit length on the centre conductor is $+q$ and on the outer $-q$. Symmetry suggests that $\mathbf{E}$ points cylindrically

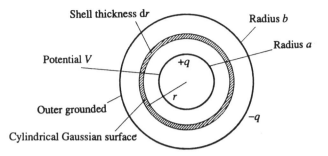

Figure 9.21   A coaxial cable with Gaussian surface at radius *r*

outwards from the centre conductor, and the lines terminate on the outer conductor. The field is clearly non-uniform.

Consider a cylindrical Gaussian surface at an arbitrary radius *r*. Assume the Gaussian surface extends for unit length into the paper.

On the curved portion of the Gaussian surface **E** is constant and parallel to the outward normal **ds** and therefore $\mathbf{E} \cdot \mathbf{ds} = E \, ds$. On the ends of the cylinder **E** is perpendicular to the normal and therefore $\mathbf{E} \cdot \mathbf{ds} = 0$ and there is no contribution. Therefore from Gauss's law

$$E \int_S ds = E2\pi r = q/\varepsilon$$

and

$$E = q/2\pi\varepsilon r \tag{9.20}$$

E is decreasing as we move outwards and therefore it is necessary to integrate to find the potential *V*. Using equation (9.9) and noting that **E** is parallel to **r**

$$V_b - V_a = -\int_a^b \mathbf{E} \cdot \mathbf{dr}$$

$$0 - V = -\int_a^b q \, dr/2\pi\varepsilon r$$

$$V = q\ln(b/a)/2\pi\varepsilon$$

and since $C = q/V$

$$C = 2\pi\varepsilon/\ln(b/a) \tag{9.21}$$

A typical coaxial cable has polythene dielectric with $\varepsilon_r = 2.2$ and dimensions $2a = 0.9$ mm and $2b = 3.3$ mm. Substituting these values in equation (9.21) gives a result for the capacitance of 94 pF/m.

9.8 Energy in the electric field

A charged capacitor stores electrical energy and may be used to do work. An uncharged capacitor has zero potential difference across it. As the capacitor is charged a potential difference develops, and positive and negative charges appear on the two plates. Note that charge is merely being redistributed and the conservation of charge is not violated. The energy in the capacitor may be calculated by finding how much work is necessary to separate the charges.

Consider a parallel plate capacitor of capacitance C that is partially charged to a value q'. It is intended to charge the capacitance to a final level q. Values on the diagram in Figure 9.22 are all dashed to show intermediate values.

Let us move a small charge dq' from the bottom to the top plate. The force acting against us due to the field E' is $dq'\,E'$. Therefore the work done is $dq'\,E'd$ which may be written as $V'\,dq$. Remembering that $q' = CV'$, the total work done to charge the capacitor from a value $q' = 0$ to $q' = q$ is

$$W = \int_0^q V'\,dq' = \int_0^q q'\,dq'/C$$

$$= q^2/2C \tag{9.22}$$

This is an important relationship; using the equation $q = CV$ it may also be expressed as $CV^2/2$ or $qV/2$. Note how similar these expressions are to kinetic energy in mechanics, which has the value $mv^2/2$.

The reason for the presence of the factor $1/2$ in all these expressions is due to *averaging*. In building up the energy in the capacitor work is done *on the average* against one-half of the final potential value.

To appreciate the values of the amount of stored charge, consider some typical figures: when $C = 100\ \mu F$ and $V = 100\ V$, then the energy stored is 0.5 J.

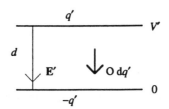

Figure 9.22 An intermediate condition in the charging process

___ **Example 9.11** _____

An electron is placed in a uniform field of 10 000 V/m. An external agent is then used to push the electron against the field for a distance of 20 cm. What is the work done against the field?

The force exerted by the field on the electron is

$$F = eE = (1.6.10^{-15})10^4 = 1.6.10^{-15} \text{ N}$$

The work done to move 20 cm is given by

$$W = eEd = (1.6.10^{-15})0.2 = 3.2.10^{-16} \text{ J}$$

It may be noted that the expression $W = eEd$ can be written as $W = eV$ where V is the potential difference across 20 cm of field.

In this case no factor of $1/2$ appears since the opposing field is constant throughout the action.

___ **Example 9.12** _____

A parallel plate capacitor with air dielectric has dimensions $A = 10 \text{ m}^2$ and $d = 0.1$ mm. The air has a breakdown field $E_{bf} = 3.10^6$ V/m. Determine the maximum voltage that can be applied to the capacitor and the energy stored at this voltage.

The maximum voltage is

$$E_{bf}d = 3.10^6(10^{-4}) = 300 \text{ V}$$

The capacitance is

$$\varepsilon_0 A/d = 8.85.10^{-12}(10)/10^{-4} = 0.885 \text{ μF}$$

The energy stored is

$$CV^2/2 = 0.885.10^{-6}(9.10^4)/2 = 0.04 \text{ J}$$

All dielectrics have a breakdown field, and therefore all practical capacitors have a maximum specified voltage rating to ensure satisfactory operation.

9.8.1 Energy density

The energy density or energy per unit volume can be derived by dividing the capacitor energy by the volume of the capacitor (see the last example). For a capacitor with plate area A, plate separation d, and dielectric constant ε we know that $C = \varepsilon A/d$ and the capacitor volume is Ad. Substituting for C in the expression $CV^2/2$ and dividing by Ad we obtain

$$\text{Energy density} = \varepsilon E^2/2 \qquad (9.23)$$

The units of energy density are J/m^3.

Equation (9.23) has been derived using an example where the field is uniform. If the field changes in a given volume V then this must be allowed for. Then the total energy in the volume is found by integrating the energy over all the small regions dv making up the volume V. When E is the field at the small volume dv then

$$\text{Total energy} = \int_V \varepsilon E^2 \, dv/2 \qquad (9.24)$$

For the previous example 9.12 the energy density is 40 J/m^3. This can be found from equation (9.23) or the total energy and the dimensions of the capacitor.

9.9 Electric field and conductors

In the presence of conductors the electric field produces a flow of charge or a current. In circuits current refers to the flow in wires and components. In fields it is more appropriate to use *current density*, for the current flow patterns are required in more complex three-dimensional conducting bodies.

9.9.1 Current density

Consider the simple conductor shown in Figure 9.23.

I is the current flowing in the conductor. $\mathbf{J}$ is defined as the current density or amps per unit area in the conductor. The quantity $\mathbf{J}$ has direction and therefore must be treated as a vector. The relationship between them if the current density is constant across the conductor is, in scalar form

$$I = JA \qquad (9.25)$$

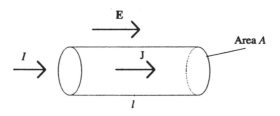

Figure 9.23 A section of a cylindrical ohmic conductor

In the general case $\mathbf{J}$ may vary over the cross-section and the cross-sectional area may not be at right angles to $\mathbf{J}$. It is then necessary to consider a small area ds and integrate over the total cross-section. The scalar product is needed to ensure that $\mathbf{J}$ is resolved along the $\mathbf{ds}$ direction. The current is then given by

$$I = \int_S \mathbf{J} \cdot \mathbf{ds} \qquad (9.26)$$

The direction of $\mathbf{J}$ is taken to be the direction of conventional current flow. Using the current density and the relationship already established between electric field and potential difference it is now possible to develop a field form for Ohm's law.

9.9.2 Ohm's law

The three-dimensional form of *Ohm's law* suitable for field use is developed simply without vectors using Figure 9.23 and well-known circuit relationships. The voltage across the conductor is $V = El$, using equation (9.18). The current through the conductor is $I = JA$ using equation (9.25).

Therefore using Ohm's law

$$R = V/I = El/JA$$

Now $R = l/\sigma A$ where σ is the conductivity. *Conductivity* σ is the inverse of ρ, the resistivity, and its units are siemens/m. The above expression may be written as

$$l/\sigma A = El/JA$$

Therefore

$$J = \sigma E$$

Current flows in the direction of the field, and so adding vector notation

$$\mathbf{J} = \sigma \mathbf{E} \qquad (9.27)$$

As expected this field description of Ohm's law uses quantities $\mathbf{J}$ and $\mathbf{E}$ instead of I and V.

In many instances in field work a *perfect conductor* with $\sigma = \infty$ is considered for simplicity. In such a conductor it is not possible to maintain a field; in circuit terms there is a short circuit. Therefore E is *always zero in a perfect conductor*.

_____ **Problems** _____

P9.1 For the arrangement of Figure P9.1 find the force acting on a unit test charge placed at P. Determine the force in both magnitude and direction.

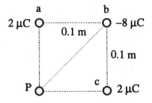

Figure P9.1

P9.2 Two equal positive point charges of $2\,\mu C$ are placed at diametrically opposite corners of a square. Two equal negative point charges are held fixed at the other two corners. It is found that no net forces act on either of the two positive charges. Find the magnitude of the negative charges.

P9.3 Two small identical conducting balls have charges of opposite polarity and different magnitude q and $-Q$. Placed a distance d from each other an attractive force F is experienced. The balls are now brought into contact and then separated to distance d again. The magnitude of the repulsive force now experienced is equal to one half of F. Find the ratio Q/q.

P9.4 Two equal positive charges are held a fixed distance 40 cm apart (Figure P9.2). A positive test charge is located in the plane perpendicular to the line joining the charges and passing through the mid-point O.

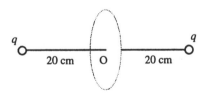

Figure P9.2

Find the radius of the circle in this plane that is the locus of points where the test charge experiences a maximum force.

P9.5 Consider the charge arrangement of problem P9.2 with the square side of length 10 cm. One of the negative charges is removed. Find the field strength and direction at the site of the removed charge.

P9.6   Using Coulomb's law for the electric field of a point charge show that the electric field at distance y from an infinite line charge of density λ C/m is given by $\lambda/2\pi\varepsilon_0 y$.
(Hint: It is probably easiest to integrate with respect to angle.)

P9.7   Using the answer of problem P9.6, find the electric field at a distance y above an infinite sheet of charge of density σ C/m^2.

P9.8   For the charge arrangement of Figure P9.3 find the position(s) on the line OP where the potential is zero.

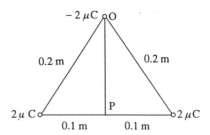

Figure P9.3

By considering the potential at other 'simple' points on the triangle draw a rough sketch for the equipotential lines of this system of point charges.

P9.9   A uniform electric field of value 50 V/m acts along the x direction.
Find the potential difference between points A and B in Figure P9.4 by utilizing the path shown between A and B.

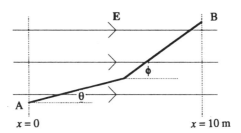

Figure P9.4

What do you conclude about the path and the potential difference?

P9.10   An electrostatic dipole consists of two equal and opposite charges a fixed short distance $2d$ apart. Show that the potential at P in Figure P9.5 due to an electrostatic dipole may be written as $p \sin \theta/4\pi\varepsilon_0 r^2$ if $r \gg d$. p is called the *dipole moment* and has the value $2qd$. Hence show that the field $E = p/4\pi\varepsilon_0 r^3$ when $\theta = 0$ and $r \gg d$.

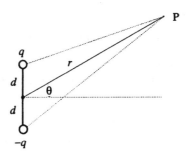

Figure P9.5

P9.11 Find the work done to move a positive charge of $10\,\mu C$ from infinity to a point 0.5 m from a fixed positive charge of $10\,\mu C$. Assume there are no other charges present.

P9.12 Two charges of magnitude 1 C and opposite polarity are placed 10 m apart.

    (a) Determine the force between the charges.
    (b) Determine the potential 1 m from the positive charge.
    (c) Determine the potential energy of the system.
    (d) Equipotentials are perpendicular to field lines. Why is this? Sketch lines of each type for this arrangement of charges.

P9.13 Charges are placed in the xy plane, 3 C at point $(0, 1)$ and -1 C at point $(0, -1)$. Find the locus of all points in the xy plane where the potential is zero.
    If you cannot do this analytically, do it by construction and plotting.

P9.14 Find the electric field at a distance y above an infinite sheet of charge of density σ C/m^2 using Gauss's law. How does this method compare with that of problem P9.7?

P9.15 A coaxial cable has a dielectric with relative permittivity $\varepsilon_r = 3$, and is charged with $+2Q$ C/m on the inner conductor and $-5Q$ C/m on the outer conductor. Find the electric field strength in the dielectric and outside the cable.
    If the cable has an inner conductor of radius 1 mm, an outer conductor of radius 3 mm, and $Q = 0.4\,\mu C$, determine the maximum field strength in the dielectric. Sketch the electric field lines for the cable.

P9.16 A parallel plate capacitor containing a dielectric with $\varepsilon_r = 10$ has plate area $A = 1$ m², and plate separation $d = 0.1$ mm. The charge on the capacitor is 0.001 C. Using Gauss's law determine the field inside the capacitor.

Deduce also the voltage across the capacitor, and the capacitance.

P9.17 A capacitor is made from two thin concentric conducting spherical shells with radii a and b where $a > b$. A charge Q is put on the capacitor. Find the capacitance of the arrangement.

(Hint: Use Gauss's law.)

P9.18 Calculate the capacitance of a parallel plate capacitor that consists of two square metal plates, each of side 50 cm, separated by a dielectric 1.5 mm thick and of relative permittivity 4. If a potential difference of 1000 V is applied between the plates, calculate the charge on the capacitor, the electric field strength and the electrical energy density.

P9.19 A parallel plate capacitor has two plates of area 0.5 m² each, spaced 2.5 mm apart. A dielectric of relative permittivity 3 and thickness 2 mm is inserted between the plates. Calculate the capacitance of the capacitor thus formed.

If a potential difference of 500 V is applied across the capacitor find

(a) the electric field strength in the dielectric and the air,
(b) the energy stored in the capacitor.

P9.20 A slab of insulating material 1 mm thick is inserted between the plates of a parallel plate capacitor. It is found necessary to increase the spacing between the plates by 0.8 mm to restore the capacitance to its original value. Calculate the relative permittivity of the slab.

P9.21 A parallel plate capacitor with a plate separation of 0.2 cm is immersed in a liquid dielectric with relative permittivity 20. The plates are charged to a potential difference of 20 kV. Find

(a) the energy/cm² stored in the capacitor,
(b) the force/cm² between the plates.

(Hint: (b) is a more difficult question – to find the force, isolate the charged capacitor (why?) and move one plate by a small amount.)

P9.22 In problem P9.20, a charge of 0.01 C is placed on the capacitor initially, before the slab is inserted. For this problem the spacing of the plates at this time is 2 mm. Find the energy stored per unit area of the plates

(a) initially,
(b) after the the slab has been inserted,

(c) after the plate spacing has been increased to restore the capacitance value.

Comment critically on the three results.

P9.23 An electron is projected at a velocity of 5×10^7 m/s into an electric field of value 75 kV/m. The field directly opposes the motion of the electron. Determine how far the electron will travel before coming to rest. (You may assume that $m_e = 9.1 \times 10^{-31}$ kg and $e = 1.6 \times 10^{-19}$ C.)

P9.24 In an electrostatic CRT an electron is accelerated from rest by a potential of 400 V, and then enters an electric field E of value 10 kV/m at right angles to its path. How much will the electron be deflected after moving through a distance of 10 cm in the field E?

P9.25 An electron revolves with a circular trajectory of constant radius in the space between two coaxial cylinders carrying charges per unit length of $+\lambda$ on the inner cylinder and $-\lambda$ on the outer.

(a) Determine an expression for the velocity of the electron.
(b) What is the velocity when $\lambda = 5 \times 10^{-8}$ C/m?
(c) What will be the effect if another electric field acting along the direction of the cylinder axis is also applied?

CHAPTER TEN

Magnetic fields

Permanent magnets and current-carrying conductors produce a magnetic field. This is easily shown using magnetic compass needles or iron filings that line up with the field and imply the presence of forces in the space around the magnet or current.

A 'magnetic flux' in space is postulated to account for the phenomenon. The density of this flux is of primary importance in the study of magnetic fields. It is found that the magnitude of this flux density is strongly affected by the presence of ferromagnetic materials such as iron, nickel, and cobalt. Such materials therefore are intimately involved when designing efficient magnetic circuits.

It is also found that electric charges experience forces in magnetic fields. These forces are more complex than the force on charges due to the electric field since the velocity of the charged particle is involved. The following section uses this last property to define the flux density of the field.

10.1 Magnetic flux density

The magnetic flux density **B** may be defined in terms of the force it produces on an isolated electric charge. Unlike electrostatic forces **B** only produces a force on a charge when the charge is *moving*. Consider the situation in Figure 10.1. The crosses represent magnetic flux lines into the paper. A positive charge q moves from left to right through the flux at a velocity v. B and v have magnitude and direction and therefore are completely described by vectors **B** and **v**.

Experimentally it is found that a force acts *upwards* on the charge at *right angles* to the velocity and magnetic flux. Also it is observed that the force is proportional to q, v, and B. Therefore an expression for the force for the situation in Figure 10.1 is $F = k(qvB)$ where k is a constant. The

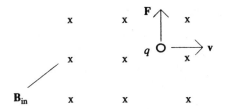

Figure 10.1 The force on a moving charge in a magnetic field

units of B are chosen so that the constant k is unity; the unit is called the *tesla*.

A magnetic flux density of 1 tesla exists when a force of 1 newton is caused by a point charge of 1 coulomb passing at a velocity of 1 m/s through the flux and at right angles to it.

Furthermore if the velocity direction is changed so that it acts at an angle θ to the magnetic flux then the magnitude of the force is found to be proportional to $\sin \theta$. Therefore the final expression for the force is $qvB \sin \theta$, and the direction of this force is always at right angles to both v and B. The force, in both magnitude and direction, can be neatly expressed using a vector product:

$$F = q(v \times B) \tag{10.1}$$

The force F has a magnitude $qvB \sin \theta$ and a direction perpendicular to both v and B such that a right-handed screw from v to B advances along F when q is positive.

10.1.1 Magnetic flux

The *magnetic flux* is a *scalar* since total amounts of flux irrespective of direction are usually considered; it is given the symbol Φ and the unit of magnetic flux is the *weber*.

The concept of flux and flux density is developed in Chapter 9; the same concept is true for magnetic fields. Therefore the relationship between flux and flux density is given by

$$\Phi = \int_S B \cdot ds \tag{10.2}$$

where S is the area of the surface across which the flux passes.

Webers are therefore equivalent to teslas m^2. In old textbooks webers/m^2 were used instead of teslas as the units of flux density.

10.1.2 The Lorentz relation

Electric and magnetic forces on a charge q can be present simultaneously and often are in electronic systems. Under these circumstances the expression for the force on the charge is obtained by superposing equations (9.4) and (10.1):

$$\mathbf{F} = q(\mathbf{E} + \mathbf{v} \times \mathbf{B}) \tag{10.3}$$

This equation is known as the *Lorentz relation*.

_____ **Example 10.1** _____

An electron travelling at velocity $v = 10^6$ m/s enters a region in which magnetic field and electric field coexist. E is produced by two plates 2 cm apart and with a potential of 100 V applied. Find the direction and smallest value for B such that the electron suffers no deflection.

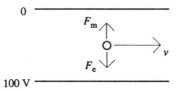

Figure 10.2

In Figure 10.2 F_e and F_m are the forces due to the electric and magnetic fields respectively.

The electric field is

$$V/d = 100/0.02 = 5000 \text{ V/m}$$

The smallest value for B will occur when $\mathbf{v}$ and $\mathbf{B}$ are at right angles. For no deflection $F_e = F_m$ and when $\mathbf{v}$ and $\mathbf{B}$ are at right angles

$$|\mathbf{v} \times \mathbf{B}| = vB$$

Therefore in magnitude $E = vB$ since q is a common quantity.

The magnitude of B is $5000/10^6 = 0.005$ T. The direction of F_e is vertically down since the electron has negative charge. Therefore the magnetic force F_m must be vertically up. For $e(\mathbf{v} \times \mathbf{B})$ to be up, B must be out of the paper.

_____ **Example 10.2** _____

An electron is shot at $v = 10^6$ m/s into a region where a magnetic field of magnitude 0.001 T is switched on as shown in Figure 10.3. What is the subsequent motion of the electron?

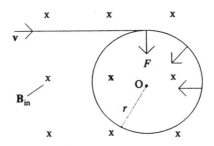

Figure 10.3

The mass of an electron is 9.1×10^{-31} kg and its charge is -1.6×10^{-19} C.

A force of magnitude $F = evB$ acts perpendicular to both v and B as indicated. This force does not affect the velocity but will bend the trajectory of the electron down. Since the force is perpendicular to the velocity at all times, the direction of the force will vary as shown. The result is that the electron will follow a circular trajectory of radius r. The value of r can be obtained by equating the magnetic force to a centrifugal force of value mv^2/r. Therefore

$$evB = mv^2/r$$

giving

$$r = mv/eB = 0.57 \text{ cm}$$

10.2 Magnetic force due to current

Current may be regarded as an assembly of charges moving at constant velocity. In reality these charges are electrons, but it is common practice to consider a conventional current flow to avoid the negative signs. For this discussion assume positive carriers moving from left to right.

Consider a small portion of a conductor with the dimensions of Figure 10.4. Let v_d be the drift velocity of the charge carriers, n the number of carriers per unit volume, and q the charge on each carrier.

The number of carriers passing a cross-section at Q per second is the number in a volume Av_d, which is $v_d An$. Therefore the current is

$$\mathbf{I} = \mathbf{v}_d Anq \tag{10.4}$$

The conductor is now put in a magnetic field with flux density B as in Figure 10.5. Each charge experiences a force given by equation (10.1)

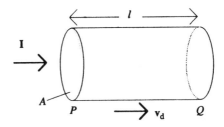

Figure 10.4   Current carriers in a cylindrical conductor

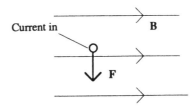

Figure 10.5   The force on a current element in a magnetic field **B**

and the charge in the conductor is $qlAn$. Therefore the total force

$$\mathbf{F} = (qlAn)\mathbf{v}_d \times \mathbf{B} \tag{10.5}$$

Using equation (10.4)

$$\mathbf{F} = (\mathbf{l} \times \mathbf{B})I \tag{10.6}$$

In this expression the vector sign has been accorded to the length l instead of I. This is normal; $\mathbf{l}$ and $\mathbf{I}$ always have the same direction. Referring to Figure 10.5 the vector product tells us that for conventional current flow the force on the current element is down, and of magnitude lBI when $\mathbf{l}$ and $\mathbf{B}$ are at right angles. This is an important relationship since it is the basis for forces in rotational machines, as seen in the next example.

_____ **Example 10.3** _____

A loop of wire of 50 turns carrying 2 A is placed in a magnetic field as shown in Figure 10.6; $a = 5$ cm, $b = 10$ cm, and the flux density is 0.01 T. Demonstrate that a torque will act on the loop which will turn the loop until it is in a vertical position. What is the maximum value of the torque?

Sections AC and A'C' act against each other and therefore have no overall effect on the motion of the loop which is constrained by the axis OO'.

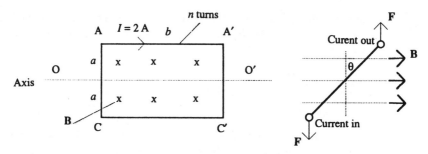

Figure 10.6

Sides AA' and CC' have an effect as shown by the forces F:

Force due to AA' $\mathbf{F} = nI(\mathbf{b} \times \mathbf{B})$ always upwards
Force due to CC' $\mathbf{F} = nI(\mathbf{b} \times \mathbf{B})$ always downwards
The torque magnitude $T = 2nIbBa \sin \theta$

This has a maximum value of

$$T = 2nIbBa = 0.01 \text{ N m}$$

The forces are always in the same direction as shown and therefore when the loop reaches a vertical position it will stop moving.

10.3 Magnetic field strength

Magnetic flux is generated by currents. Experimentally it is found that the flux density dB at a given distance r from a current element dl carrying a current I is proportional to dl and I, and inversely proportional to the distance r squared. This may be written as $dB = kIdl/r^2$ when r is perpendicular to dl. In the SI system the proportionality constant k in free space has the value $\mu_0/4\pi$. μ_0 is called the *permeability of free space* and has the value given below. This value ensures that the velocity of electromagnetic waves in free space, which is given by $c = 1/(\mu_0 \varepsilon_0)^{1/2}$, is 2.998×10^8 m/s.

In a magnetic medium μ_0 is replaced by the permeability μ which embodies the effect of the material in which B is produced. Further details of the current element are given as the Biot–Savart law in section 10.6

The magnitude of the flux density $\mathbf{B}$ is determined by the currents present and the material in which the flux is generated. In some materials known as *ferromagnetics* the effect of the material is very pronounced. It is useful to define another magnetic vector that is dependent on the currents *only*. This vector is called the *magnetic field strength* and is given the symbol **H**. It may be defined simply in terms of **B** by the relationship

$$\mathbf{B} = \mu\mathbf{H} \tag{10.7}$$

where

$$\mu = \mu_r\mu_0 \tag{10.8}$$

μ_0 is the *permeability of free space* and has the value $4\pi \times 10^{-7}$ henries/m; μ_r is the relative permeability and can achieve very large values for special magnetic alloys. For materials which are *not ferromagnetic or ferrimagnetic* μ_r is approximately unity.

The SI unit for **H** is amperes/m. Note the similarity of equation (10.7) with the equivalent electrical equation (9.13). The simplicity of equation (10.7) is a little misleading and further discussion regarding magnetization and the meaning of the equation may be found in Appendix 5.

In particular if a graph is drawn of B against H for a ferromagnetic material the result is *non-linear*. More on this aspect can be found in section 10.7 and Appendix 5.

10.4 Gauss's law for magnetism

This law may be obtained by analogy with Gauss's law for electricity. This law states that the electric flux through a closed surface S is equal to the charge contained within S divided by ε. The equivalent to electrical charge is the magnetic 'charge' or *magnetic pole strength*. Most students are familiar with the concept of magnetic poles from their knowledge of permanent magnets.

Experimentally it is found that magnetic poles *always* occur as *dipoles*. Even at the atomic level, a hydrogen atom for example, the single atom acts as a small north–south dipole. Therefore since there is always the same number of north and south poles the total amount of magnetic 'charge' enclosed by any surface is *zero*. Gauss's law for magnetism may therefore be stated as:

The magnetic flux through a closed surface S is zero.

The law is written as

$$\oint_S \mathbf{B} \cdot \mathbf{ds} = 0 \tag{10.9}$$

Equation (10.9) is one of the major laws of electromagnetism.

In an introductory course most problems concentrate on the current as the source of the field, and therefore equation (10.9) is used less than the equations developed in the following sections.

10.5 Ampère's law

Experiments show that a magnetic flux will be set up around a current or currents. The flux forms closed loops around the current. In particular for the case of a single straight conductor the lines of constant flux are circles around the conductor and become weaker as the distance from the conductor is increased.

Ampère's law gives an effective method of calculating the values of the magnetic flux produced by currents whenever symmetry is present. Consider a closed contour C of any shape surrounding a current I (Figure 10.7). B may be in any direction with respect to the contour or the current.

Ampère's law states that:

The resolved value of B along the contour multiplied by the distance dl, taken round the complete contour, is equal to the total current surrounded by the contour multiplied by the permeability of free space.

This is written concisely using a scalar product as

$$\oint_C \mathbf{B} \cdot \mathbf{dl} = \mu_0 I \tag{10.10}$$

This is true whatever the shape of the contour C and for any number of currents. The contour C must be closed and surround the current(s).

If B is produced in a magnetic material then μ_0 is replaced by μ.

Although the equation is generally true, the integral becomes difficult if symmetry is not present. It is necessary to choose contours that are symmetrical with respect to the currents.

It is possible that the current within the contour may be specified as a current density which varies over the area defined by the contour. In these circumstances we can make use of equation (9.26) and write Ampère's

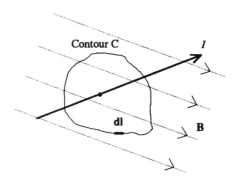

Figure 10.7 A contour C surrounding a current *I* in a field **B**

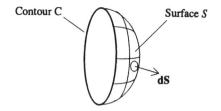

Figure 10.8 An open surface S with its defining contour C

law as

$$\oint_C \mathbf{B} \cdot \mathbf{dl} = \mu_0 \int_S \mathbf{J} \cdot \mathbf{ds} \tag{10.11}$$

where S is the *open surface* bordered by the contour C (Figure 10.8). Think of a schoolboy's cap where the contour is the rim!

10.5.1 Magnetic flux density around a straight wire

The infinitely long straight wire is important because many other current arrangements can be made up from it.

Experiment shows that the loops of flux around the wire are circular and tangential to the radius, and therefore a circular contour C is chosen (Figure 10.9). If B is desired at radius r then the contour must be placed at this radius. Since **B** is parallel to **dl** on C and **B** is constant on the contour, then

$$\oint_C \mathbf{B} \cdot \mathbf{dl} = B \int_{2\pi r} dl = B2\pi r$$

From equation (10.10)

$$B = \mu_0 I / 2\pi r \tag{10.12}$$

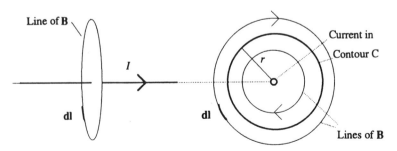

Figure 10.9 A contour C at radius r about a long straight conductor

Note that cylindrical symmetry gives 2π in the denominator. The direction of B is given by a right-handed screw advancing along the current direction. B is constant on circles about the wire.

If we go *inside* the wire with our contour then less current is enclosed and the value of B falls. Clearly at the centre of the wire $B = 0$. The value of B is greatest at the wire surface and has a value $\mu_0 I/2\pi a$ for a wire of radius a. Problem P10.7 at the end of the chapter is concerned with this feature.

_____ **Example 10.4** _____

Two infinitely long conductors each carrying 10 A in the same direction pass into the paper at X and Y as show in Figure 10.10. What is the magnetic flux density at point P?

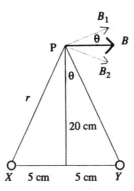

Figure 10.10

Each conductor produces a flux at P as though the other were not present. Therefore the effect of each can be found and then the resultant determined. The conductor at X produces a flux density B_2 at P, and that at Y a flux density B_1. The value of the flux densities will be given by equation (10.12) with the relevant values inserted. It is also evident since $B_1 = B_2$ that vertical components will cancel and the final resultant will point to the right and be of value

$$B = 2B_1 \cos \theta$$
$$= 2\mu_0 I \cos \theta / 2\pi r$$
$$= 1.88.10^{-5} \text{ T}$$

since $\mu_0 = 4\pi \times 10^{-7}$ H/m, $I = 10$ A, $\theta = \tan^{-1} 5/20 = 14°$, and $r = 20.62$ cm.

10.5.2 Magnetic flux density in a long solenoid

This is an important topic since solenoids are a common method of producing magnetic flux in practice. Consider an air-cored solenoid (Figure 10.11) with n turns/m and carrying a current I.

Experimentally it is found that in a *long* solenoid as in Figure 10.11 there is negligible flux outside the solenoid in the central region and a strong concentration of flux *parallel* to the solenoid axis inside. Ampère's law may be used to find a value for **B** by specifying a suitable contour. Let this contour be abcd and let us traverse the contour in direction abcd. Each side of the contour is different and must be treated separately. We have

$$\oint_C \mathbf{B} \cdot \mathbf{dl} = \int_{ab} \mathbf{B} \cdot \mathbf{dl} + \int_{bc} \mathbf{B} \cdot \mathbf{dl} + \int_{cd} \mathbf{B} \cdot \mathbf{dl} + \int_{da} \mathbf{B} \cdot \mathbf{dl}$$

$$= Bh + 0 + 0 + 0$$

B and **dl** are parallel on ab; therefore $\mathbf{B} \cdot \mathbf{dl} = B \, dl$ and B is also constant on ab. Therefore $\int B \, dl = B \int dl = Bh$. Over ad and bc, **B** and **dl** are at right angles; therefore $\mathbf{B} \cdot \mathbf{dl} = 0$. On cd $B = 0$, and therefore this portion does not contribute to the integral. The contour encircles nh turns and so by Ampère's law

$$Bh = \mu_0 I n h$$

Therefore

$$B = \mu_0 I n \tag{10.13}$$

It should be noted that this value for B is obtained wherever we place the contour section ab within the solenoid; the value for B is constant over the cross-section.

If the solenoid has a core of magnetic material with relative permeability μ_r then we merely replace μ_0 by $\mu_r \mu_0$.

This is a simple result and will be approximately true in the centre of all solenoids where the length is large compared with the radius. At the

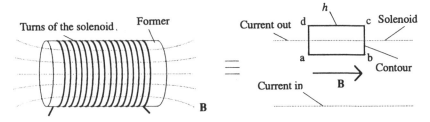

Figure 10.11   A simple contour surrounding current in a solenoid

ends of the solenoid the value of B will become smaller as the flux begins to spread.

For a solenoid with $n = 1000$, $I = 10$ A, and $\mu_r = 100$, the value for the magnetic flux density inside the solenoid is $B = 1.26$ T.

10.6 Biot–Savart's law

Ampère's law is the best method of calculating B when symmetry is present. When this is not so then Biot–Savart's law is of great value. The law is equivalent to Coulomb's law for electrostatics and deals with small elements of current.

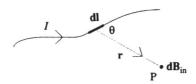

Figure 10.12 A current element **dl** produces a field **dB** into the paper at P

In Figure 10.12 consider a small element dl on the wire which carries a current I. The Biot–Savart law gives a value for the flux density at P due to an element dl. Consider point P at distance r from dl and such that **dl** and **r** make an angle θ with each other. The vector **r** is assumed to act from dl to P. *Then the magnetic flux density due to the element of current dl is given in magnitude and direction by the expression*

$$\mathbf{dB} = \mu_0 I(\mathbf{dl} \times \hat{\mathbf{r}})/4\pi r^2 \tag{10.14}$$

The direction for **dB** will be into the paper when **dl** and **r** have the directions shown. Remember $\hat{\mathbf{r}}$ has unity magnitude.

The value of **B** for any shape or length of wire can be found by integrating over the wire length. The integral is not always easy. B due to an infinite straight wire can be found by this method; example 10.6 will show how.

_____ **Example 10.5** _____

A circular current loop of radius R carries a current I (Figure 10.13). Find the magnetic flux density at P at distance x along the axis of the loop.

Consider a current element dl at the top of the loop. Since the current is out of the paper and r is pointing towards P, then the direction of the elemental dB will be as shown. **dl** and **r** are at right angles so sin θ due to

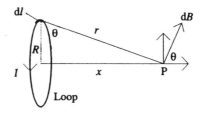

Figure 10.13

the vector product is unity. The resultant magnitude of **dB** is

$$dB = \mu_0 I \, dl/4\pi r^2$$

dB may be resolved into vertical and horizontal components. The vertical component will cancel with the flux density dB due to an element on the loop diametrically opposite, and overall there will be no vertical B.

The resultant B will be along the x-axis only, and may be found by integrating over all the current elements around the loop. For this example R, r, x, θ, are all constant, only dl is variable. From Figure 10.13 it is clear that $\cos \theta = R/r$. Therefore

$$B = \int dB \cos \theta = \int \mu_0 I \cos \theta \, dl/4\pi r^2$$

$$= (\mu_0 IR/4\pi r^3) \int_{2\pi R} dl = \mu_0 IR^2/2r^3$$

_____ **Example 10.6** _____

Calculate B at distance x from an infinitely long straight wire carrying a current I (Figure 10.14).

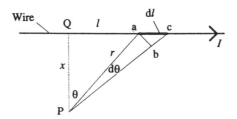

Figure 10.14

Consider a small element dl on the wire distance r from P and l from Q. On this diagram l, r, and θ are variables. It is necessary to choose which variable to use for the problem; in this case it will be found advantageous to integrate with respect to the angle θ. The angle $d\theta$ is specified as the angle turned through as l changes by dl.

Using Biot–Savart's law we have

$$\mathbf{dB} = \mu_0 I(\mathbf{dl} \times \hat{\mathbf{r}})/4\pi r^2$$

$\mathbf{dl}$ is to the right, $\mathbf{r}$ points towards P and the angle involved in the vector product is Pac which is $(90° + \theta)$. dB will therefore be into the paper and have magnitude

$$dB = \mu_0 I \sin(90° + \theta)dl/4\pi r^2$$

Now $\sin(90° + \theta) = \cos\theta$; $ab/ac = r\,d\theta/dl = \cos\theta$; $x/r = \cos\theta$.

Substituting to change from variables l, r, and θ to variable θ only,

$$dB = \mu_0 I\,d\theta/4\pi r$$
$$= \mu_0 I \cos\theta\,d\theta/4\pi x$$

Integrating

$$B = (\mu_0 I/4\pi x)\int_{-\pi/2}^{\pi/2}\cos\theta\,d\theta$$
$$= \mu_0 I/2\pi x$$

This is the same expression as determined in section 10.5.1. It should be noted that the derivation in 10.5.1 is considerably easier than the Biot–Savart approach.

10.7 Magnetic circuits

For a given current the amount of magnetic flux created can be drastically increased by using a *high permeability magnetic material*. Therefore any device requiring magnetic flux can be made more efficient by using magnetic material in, the circuit. Devices of this nature include motors, generators, and galvanometers which therefore all use magnetic material. The loop of magnetic material is used to allow an easy path for the flow of magnetic flux.

In dealing with practical magnetic circuits the non-linearity of the *BH* curve (Figure 10.15) must be taken into account. The reason for the non-linearity is the complex movement of magnetic domains within the magnetic material. These domains tend to line up in the presence of external magnetic fields. Further information is given in Appendix 5.

It is not possible to find a value for *B* knowing *H* without resorting to this curve for the particular magnetic material. We will find that *load line* techniques have to be used in association with the *BH* curve.

In magnetic circuits it is common to use Ampère's law with *H* instead

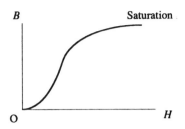

Figure 10.15   The non-linear relationship between B and H

of B, since μ can vary around the contour C. It is also common to have multiple turns N through the contour. Therefore Ampère's equation may be written as

$$\oint_C \mathbf{H} \cdot \mathbf{dl} = NI \tag{10.15}$$

10.7.1 A simple magnetic circuit

Suppose a large flux is required in the small airgap of an ammeter. Then an arrangement such as that shown in Figure 10.16 could be used. The 'iron' magnetic circuit concentrates the flux and ensures a high flux density across the airgap. The flux is generated by a coil of N turns wound on the magnetic 'core' which is often called the *armature*.

The flux is concentrated in the armature and therefore will be *constant* round the circuit. If the cross-sectional area A of the armature varies then the magnetic flux density will vary since the flux $\Phi = BA$. The flux tends to bulge out at the airgap and therefore the flux density in the gap is less than in the armature.

The sections of the complete circuit are numbered 1 to 4. The sections must be treated separately if they have different cross-sections for then B

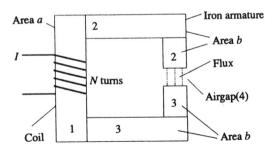

Figure 10.16   A typical magnetic circuit with an airgap

and H will change. For the circuit above 1 and 4 are treated separately but 2 and 3 can be treated as one length. Therefore

$$NI = \oint \mathbf{H} \cdot \mathbf{dl} = H_1 l_1 + H_{23} l_{23} + H_4 l_4 \tag{10.16}$$

The simplification occurs since within each section $\mathbf{H}$ is constant and parallel to $\mathbf{dl}$; hence $\int \mathbf{H} \cdot \mathbf{dl} = H \int dl = Hl$. This simplification always occurs in this type of problem. The length l used is the *mean* length of the iron path.

The quantity Hl is given the name *magnetomotive force* or *m.m.f.* Hl has units of amps or *ampere turns* (AT). An expression for m.m.f. can be obtained in terms of the flux Φ as follows:

$$\text{m.m.f.} = Hl = Bl/\mu = \Phi(l/\mu A) \tag{10.17}$$

The quantity in brackets is called the *reluctance* of the magnetic core. Therefore

$$\text{m.m.f.} = \text{flux(reluctance)} \tag{10.18}$$

It may be noted that reluctance $(l/\mu A)$ has a similar structure to resistance $(l/\sigma A)$. Two examples of magnetic circuit problems follow.

_____ **Example 10.7** _____

What current is needed to maintain a flux density of 1 T in the airgap of the circuit in Figure 10.17 given a flux leakage at the gap of 10%? All portions of the circuit have the same cross-sectional area. Values from the *BH* curve are

B (T)	0.9	1.0	1.1	1.2
H (AT/m)	900	1000	1200	145

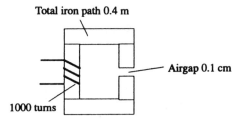

Total iron path 0.4 m

Airgap 0.1 cm

1000 turns

Figure 10.17

The flux density in the airgap is 1 T. Therefore the flux density in the iron must be 1.1 T to allow for the leakage at the gap. From the table H in iron must be 1200 AT when B is 1.1 T.

From Ampère's law

$$NI = (Hl)_{iron} + (Hl)_{air}$$
$$Hl_{iron} = 1200(0.4) \qquad = 480 \text{ AT}$$
$$Hl_{air} = 1.0(0.001)/4\pi.10^{-7} = 796 \text{ AT}$$

Therefore

$$1000I = 480 + 796$$

and

$$I = 1.276 \text{ A}$$

In calculating the AT for the airgap, the magnetic field strength H in air is found from the expression B/μ_0. The small value of μ_0 causes the airgap to have a strong effect out of all proportion to its size. In other words the airgap spoils the magnetic circuit and therefore is always made as small as possible. In galvanometers and ammeters the gap is necessary, for the rotating measuring coil must be placed in it. However, where possible, e.g. in transformers, the airgap is eliminated altogether.

_____ **Example 10.8** _____

The above example may be asked in reverse. What is the value of B in the airgap if the current in the 1000 turn coil is 1.1 A?

As before

$$NI = (Hl)_{iron} + (Hl)_{air}$$

and

$$1100 = H_{iron}(0.4) + B_{air}(0.001)/4\pi.10^{-7}$$

Now owing to the 10% leakage at the gap the flux density in the iron must be greater than in the air, i.e. $B_{iron} = 1.1B_{air}$. The equation now becomes

$$1100 = H_{iron}(0.4) + B_{iron}(0.001)/(1.1)4\pi.10^{-7}$$

This is an interesting equation. The result for B should be available to us because B and H in the iron are related. But they are related by the table of values which is non-linear. The way to deal with this is to draw a graph for the table and use a load line for the equation above. Rearranging the equation into the form $y = mx + c$ where B is the ordinate and H the abscissa gives

$$B_{iron} = -5.53.10^{-4}H_{iron} + 1.52$$

Drawing this straight line on the *BH* curve, the point of crossover gives a value $B_{iron} = 0.98$ T.

Therefore the value for $B_{air} = (0.98)/1.1 = 0.89$ T.

10.7.2 An electrical analogy

The understanding of magnetic circuits is helped by an electrical analogy. Let us write down the terms above and compare them with similar electrical terms:

Magnetic	*Electric*
m.m.f.	e.m.f.
Flux	Current
Reluctance	Resistance

The battery in an electrical circuit is mirrored by *NI*, the generating ampere turns, or 'magnetic battery'. The electrical equivalent of equation (10.18) has the form e.m.f. = current(resistance); in other words, equation (10.18) is a 'magnetic' Ohm's law. Using these analogies Kirchhoff's laws may be applied to the magnetic circuit just as in the electrical case.

10.7.3 A more complex magnetic circuit

Figure 10.18 shows a magnetic circuit and its electrical counterpart. It is with these more complex circuits that the electrical analogy is most useful. Kirchhoff's laws may be applied as in the electrical case.

R_1–R_5 are the reluctances of the various sections of the circuit. *NI* is the equivalent 'magnetic' battery. The flux Φ created by the coil on the centre limb splits into flux to the left Φ_1, and flux to the right Φ_2. The size of these fluxes depends on the reluctance values. Kirchhoff's laws

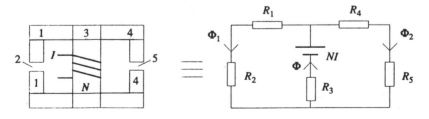

Figure 10.18 The electrical model for a magnetic circuit with two airgaps

enable us to write

$$\Phi = \Phi_1 + \Phi_2$$
$$NI = (Hl)_1 + (Hl)_2 + (Hl)_3$$
$$= (Hl)_4 + (Hl)_5 + (Hl)_3$$

Problems can be solved using these three equations and incorporating the information concerning the magnetic material. This information will normally be given as a graph of B against H.

The above description suggests that the engineering side of the problem is not difficult. However, the presence of the non-linear relationships and the number of equations make the problem more difficult than its electrical counterpart. Problem P10.18 at the end of the chapter is an exercise with such a magnetic circuit.

10.7.4 Eddy currents

The materials from which magnetic circuits are made are usually also good electrical conductors. In the presence of changing magnetic fields, due to either alternating flux or conductor movements, currents tend to be induced in the magnetic material. These currents are called *eddy currents* and cause power losses.

To reduce these *eddy current losses* magnetic circuits are *laminated*. That is, they are built up from sheets of magnetic material interleaved with thin sheets of electrical insulating material. The magnetic material is continuous in the direction of the flux apart from designed air gaps. The whole assembly of laminations is bolted tightly together to form a rigid magnetic circuit.

—————— **Problems** ——————————————————————————————

P10.1   A straight wire makes an angle of 60° to a uniform magnetic field with $B = 1$ T. If the current in the wire is 5 A, calculate the magnitude and direction of the force on the wire per unit length.

P10.2   An electron is shot at constant velocity along the y-axis of a Cartesian coordinate system into a region where a magnetic field of magnitude 0.8 T and an electric field of magnitude 1200 V/m coexist. What must be the field directions of E and B and the minimum velocity of the electron so that the electron keeps its original velocity and is not deflected?

P10.3   The length of wire on a square coil of n turns carrying a current I is L and is fixed. Show that the torque on the coil in a field B is a maximum when the coil has one turn. Find a value for the torque when $B = 0.05$ T, $I = 40$ A and $L = 24$ cm.

P10.4 A conducting rod of mass 0.2 kg and length 0.5 m is placed in a magnetic field with $B = 0.25$ T as shown in Figure P10.1. Current is passed through the rod until it just floats in the earth's gravitational field.

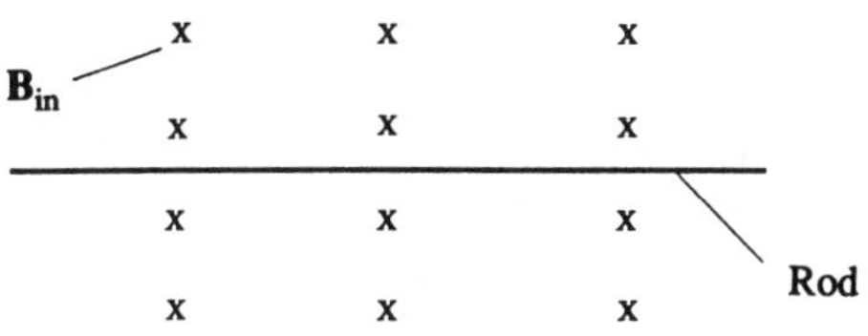

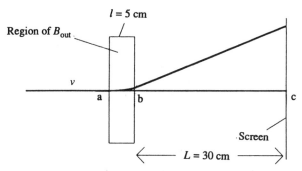

Figure P10.1

Determine the magnitude and direction of the current.

P10.5 A television CRT uses magnetic deflection as shown in Figure P10.2. An electron is accelerated to a velocity v by an attracting voltage $V = 1000$ V and then enters the magnetic field.

Figure P10.2

If $B = 0.001$ T in a direction out of the paper determine the deflections at b and c.

P10.6 The values for B and H given below are for a sample of silicon steel:

B (T)	0.075	0.2	0.4	0.6	0.75	1.00	1.15	1.2	1.23
H (A/m)	25	50	75	100	125	200	300	400	500

Plot a graph of relative permeability μ_r against H.

P10.7 A thick wire of radius R carries a current I that is uniformly distributed over the cross-section of the wire. Determine the magnetic flux density B inside and outside the wire. Sketch a graph of B against radial distance from the centre of the wire.

P10.8 Show that the magnetic field strength at distance r from a thin infinite wire carrying a current I is given by $H = I/2\pi r$.

An infinite strip of copper of width $2b$ and negligible thickness carries a current I. Using the above result determine the field of the strip in the plane of the strip as a function of the distance x ($>b$) from the centre of the strip.

P10.9   Two parallel conducting wires are placed a distance $b = 0.2$ m apart. The wires carry equal currents $I = 20$ A, but in opposite directions.

Find the value of B on a line perpendicular to the two wires, and at a distance $x = 0.05$ m from one of the wires, both between the wires and outside them.

P10.10   For the conductors of problem P10.9, find B at point P distance $x = 0.05$ m from the line joining the wires and equidistant from the two wires (Figure P10.3).

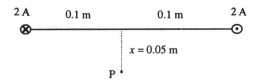

Figure P10.3

P10.11   Show that the magnetic flux density B inside a toroid is given by $B = \mu_0 In/2\pi r$, where n is the total number of turns on the toroid, I is the current in the toroid, and r is the radius from the centre of the toroid. Why is the result independent of the cross-sectional shape of the toroid?

P10.12   A toroid with 250 turns and a square cross-section has an internal radius of 80 cm and an external radius of 120 cm. The toroid carries a current of 10 A. Find

(a)   the value for B at radius 1 m,
(b)   the total flux Φ over the cross-section of the toroid.

(Note: A toroid is a long solenoid bent into a closed loop – in this case a circle.)

P10.13   Two coils, each of 500 turns, are arranged as shown in Figure P10.4. The radius of each coil is R and they are separated by a distance d. Each coil carries the same current I in the same direction. Find an expression for the value for B, on the axis, between the coils.

Determine the value of B at P, distance 1 cm from one of the coils, when $R = 5$ cm, $d = 4$ cm, and $I = 25$ A.

If d is made equal to R, then the value of B on the axis between the coils tends to be uniform. Investigate this situation.

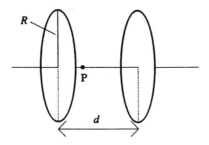

Figure P10.4

P10.14 Find the value for the magnetic flux density at a distance of 10 cm from the centre of a straight wire of length 50 cm. The current in the wire is 2.5 A.

P10.15 A square loop of wire with 100 turns carries a total length of 10 m of wire. The current flowing in the wire is 40 mA. What is the value of the magnetic flux density at the centre of the loop?

P10.16 A magnetic circuit consists of a ring of iron of mean diameter 15 cm with a single airgap of length 0.1 cm. The cross-section of the ring is 4 cm². The BH values for the iron are:

H (A/m)	40	80	120	160	800	1600	3200
B (T)	0.37	0.72	0.92	1.04	1.40	1.47	1.55

A coil of 400 turns is wound uniformly on the ring. Determine the current in the coil required to give a flux of 0.28 mWb in the airgap.

P10.17 A ring of steel has a uniform cross-section and a mean perimeter of 45 cm. An airgap of 0.05 cm is cut in the ring. The ring is wound with a coil of 500 turns. Fringing may be allowed for by taking the effective cross-section of the gap to be 5% greater than that of the steel ring. Determine:

(a) the current to give $B = 1.1$ T in the steel,
(b) the flux density in the steel if the current in the coil is 0.5 A.

You may assume the following BH values:

H (A/m)	100	150	200	350	560	750
B (T)	0.3	0.52	0.75	1.00	1.10	1.12

P10.18 The core shown in Figure P10.5 has a centre limb with a cross-sectional area of 16 cm². All other limbs have a cross-sectional area of 10 cm². A coil with 450 turns is wound on the centre limb.

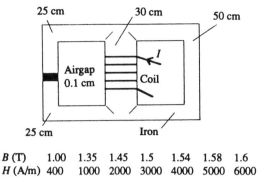

B (T)	1.00	1.35	1.45	1.5	1.54	1.58	1.6
H (A/m)	400	1000	2000	3000	4000	5000	6000

Figure P10.5

Estimate the current in the coil to give a flux density of 1.1 T in the airgap. Flux leakage at the gap may be neglected.

CHAPTER ELEVEN

Electromagnetic induction

Electromagnetic induction is the study of the situation where electric and magnetic phenomena occur together and interact. The requirement for this to happen is that there must be a *time* rate of change associated with either the electric or the magnetic fields. The rate of change may be produced by either variation of the fields in time, or movement of circuit components.

Faraday was the major influence in establishing how electric fields are produced from changing magnetic fields. He performed many interesting and penetrating experiments on electromagnetic induction at the beginning of the nineteenth century. The result of this pioneering work is *Faraday's law of electromagnetic induction.*

Maxwell, using theoretical methods, established how magnetic fields can be produced from changing electric fields with his discussion on *displacement current.*

11.1 Faraday's and Lenz's laws

Faraday's law of electromagnetic induction states:

An e.m.f. or current is induced in a closed circuit when the magnetic flux cutting the circuit is changing.

If e is the e.m.f. and Φ the flux then the equation may be written in its simplest form as

$$e = -d\Phi/dt \qquad (11.1)$$

If the circuit is a coil of N turns and the flux cuts all turns, then the equation is written as

$$e = -d(N\Phi)/dt = -N \, d\Phi/dt \qquad (11.2)$$

The quantity $N\Phi$ is often called *flux linkages*. A minus sign is included on the right hand side of equations (11.1) and (11.2) to imply a direction for the e.m.f. that is given by the law due to Lenz:

The induced e.m.f. or current is in a direction to oppose the change causing it.

Faraday's law is of prime importance in electrical engineering since it is the basis of electrical machines associated with changing magnetic flux.

_____ **Example 11.1** _____

A long solenoid with a cross-sectional area of 0.001 m^2 generates a flux density of 0.25 T inside the solenoid. A small search coil with 200 turns is placed around the solenoid and connected in series with a resistor of value $50\ \Omega$ (Figure 11.1).

(a)   If B is reversed what magnitude of charge flows round the circuit?
(b)   If B is reversed in 1 s what magnitude of current flows?
(c)   If B is reversed in 1 ms what magnitude of current flows?

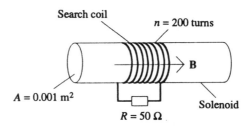

Figure 11.1

(a)   Suppose an e.m.f. e is induced in the search coil which causes a current to flow. Then the following equations hold, using $\Delta/\Delta t$ for d/dt to indicate a single change due to reversal:

$$e = iR = -n\,d\Phi/dt = -n\,\Delta\Phi/\Delta t$$

Φ can be replaced by BA, and we also have $i = \Delta q/\Delta t$.
Therefore, rearranging:

$$\Delta q = i\,\Delta t = -n(\Delta B)A/R$$

ΔB is the change in B which is 0.5 T. Therefore the magnitude of Δq is

$$\Delta q = 200(0.5)(0.001)/50 = 0.002 \text{ C}$$

(b) $\quad i = \Delta q/\Delta t = 0.002/1 \quad = 0.002 \text{ A}$

(c) $\quad i = 0.002/0.001 \quad\quad = 2 \text{ A}$

11.1.1 Field generalization of Faraday's law

The law of equation (11.1) is written in circuit terms. A field description requires the law to be restated using quantities **E** and **B**. The relationship between e.m.f. and electric field in the case of e.m.f. produced by changing flux is given by

$$e = \oint_C \mathbf{E} \cdot \mathbf{dl} \tag{11.3}$$

The relationship between flux and flux density is given in equation (10.1) and restated here as $\Phi = \int_s \mathbf{B} \cdot \mathbf{ds}$.

Therefore Faraday's law in terms of **E** and **B** is

$$\oint_C \mathbf{E} \cdot \mathbf{dl} = -d/dt \int_s (\mathbf{B} \cdot \mathbf{ds}) \tag{11.4}$$

where the contour and surface are related as in Figure 11.2. Note that the contour C must be closed to define the surface S.

In Chapter 9 it was found that the electric field $\mathbf{E}_C$, produced by charges, has the property that $\oint \mathbf{E}_C \cdot \mathbf{dl} = 0$, and that the potential difference calculated from $\mathbf{E}_C$ is independent of the path taken. A field with these properties is called a *conservative field*.

Equation (11.3) implies that $\oint \mathbf{E} \cdot \mathbf{dl}$ has a value – the e.m.f. round the closed loop. In addition the e.m.f. developed by changing flux is dependent on the path taken. Clearly the two electric fields have different properties due to the way in which they are produced. The electric field specified in equation (11.3) and produced by changing flux is often called an *e.m.f.-producing field*.

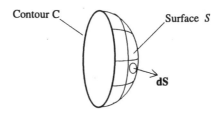

Figure 11.2 Open surface S is bounded by contour C

_____ **Example 11.2** _____

An infinitely long wire carries a current $I = 100$ A. Find the flux Φ crossing the area ABCD shown in Figure 11.3.

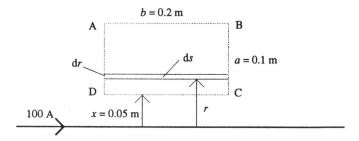

Figure 11.3

The magnetic flux density at radius r from the wire is given by the expression $B = \mu_0 I / 2\pi r$. To find the flux it is necessary to define a suitable small area $\mathrm{d}s$ and then integrate over area ABCD. In this case a strip $\mathrm{d}r$ wide at radius r is the obvious choice. This strip has an area $\mathrm{d}s = b\,\mathrm{d}r$.

B is in the form of circles around the wire, and therefore it crosses area ABCD at right angles and is out of the page. $\mathbf{B}$ and $\mathbf{ds}$ are parallel and therefore $\int \mathbf{B} \cdot \mathbf{ds}$ is equal to $\int B\,\mathrm{d}s$.

The total flux Φ across area ABCD is

$$\int B\,\mathrm{d}s = \int_{0.05}^{0.15} Bb\,\mathrm{d}r$$

$$= (\mu_0 I b / 2\pi) \int_{0.05}^{0.15} \mathrm{d}r/r$$

$$= (\mu_0 I b / 2\pi)[\ln r]_{0.05}^{0.15}$$

Substituting values

$$\Phi = 4.39\ \mu\text{Wb}$$

_____

11.2 Time-varying magnetic fields

There are two basic ways in which a rate of change of flux can be caused:

1. *Alternating currents create alternating flux.* Φ and B are constantly changing in a sinusoidal manner and may be dealt with using Faraday's law. This method of developing an e.m.f. when the circuit conductors are fixed is called *transformer induction*.

2. *Motion of circuit conductors with respect to a constant field.* In a practical situation the motion is usually obtained by the rotation of coils. This method of developing an e.m.f. is called *motional e.m.f.* Lenz's law implies that the mover will experience a force *against* its action.

Motional e.m.f. is usually treated by utilizing the magnetic part of the Lorentz law.

11.2.1 Transformer induction

There is no conductor motion and the situation is easily handled using equation (11.2) or (11.4). The main features of this type of problem are shown with reference to the interesting practical application of a toroidal current transformer. The application is to obtain an estimate of a large sinusoidal alternating current by measurement of a much smaller current. A toroid is a long solenoid bent round into a closed loop form, in this case circular.

Consider a toroid with the dimensions of Figure 11.4 wound on a ferromagnetic core whose relative permeability is assumed to have a constant value $\mu_r = 200$. A conductor carrying a large alternating current, with peak value I, passes through the toroid. A value for the current I is to be found by measuring the current developed in the toroid circuit. In order to do this it is first necessary to find the flux in the toroid.

To find B

The toroid is a magnetic circuit where flux flows owing to the enclosed current consisting of a single turn $N = 1$ carrying the current I. The length of the toroid is $l = \pi D$. For peak values

$$NI = Hl = Bl/\mu_0\mu_r$$

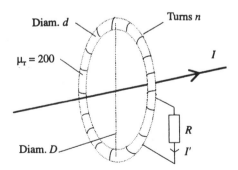

Figure 11.4 A toroid used as a current transformer

The flux density is alternating due to I and can be written as $B \sin \omega t$, where the peak value is $B = \mu_0 \mu_r I / l$.

To find e

$A = \pi d^2 / 4$ is the cross-sectional area of the toroid. For this problem $D \gg d$ and it is a good approximation to assume $\phi = BA$. Using Faraday's law for the toroid

$$e = -n \, d\Phi/dt = -nBA\omega \cos \omega t$$

If e is applied across a resistance R then a current with peak value I' will result. This current can be small and is used to estimate the large current I.

Let us consider some typical values: $d = 0.01$ m, $D = 0.05$ m, $I = 250$ A peak value at a frequency of 50 Hz, $R = 100 \, \Omega$, and $n = 240$. Inserting these values the resulting peak values are $B = 0.4$ T and $I' = 23.7$ mA.

In this situation a measurement at a level of 23.7 mA is made as a consequence of a current of 250 A, and this is why the device is called a current transformer. The ratio between the two currents is maintained provided the toroid is utilized over the linear portion of the BH curve and is not taken into the saturation region. Therefore it is possible to calibrate the device so that the large current value can be determined directly knowing the value of the toroid current.

It is also possible to design the toroid so that it is hinged on one side. It can then be placed around the large conductor and the current I determined *without breaking the circuit*.

Example 11.3

A rectangular coil with n turns and area A is situated at right angles to a field with a changing flux density given by $B = B' \sin \omega t$. Deduce an expression for the e.m.f. induced in the loop. For the values given estimate the magnitude of the e.m.f. $A = 50$ cm^2, $n = 100$, $B' = 0.1$ T, and $\omega = 300$ rad/s.

$$\text{Flux linkages} = nAB \qquad\qquad = nAB' \sin \omega t$$
$$\text{Therefore } e = -d/dt(nAB' \sin \omega t) = -(nAB'\omega)\cos \omega t$$
$$= -15 \cos \omega t$$

Clearly the e.m.f. is sinusoidal with a peak value of 15 V and 90° out of phase with the variation of B. The e.m.f. appears at the terminals of the coil.

11.2.2 Motional e.m.f.

It is found experimentally that an e.m.f. appears across a wire moving through a magnetic field. Consider a wire conductor of length l moving at a velocity v through a field of infinite extent as shown in Figure 11.5. It is desired to know the e.m.f. developed across the ends of the conductor. Let dl be an infinitesimal length on the wire.

Charges exist in the wire and must move with it. The charges are therefore subject to a force given by Lorentz's law. For a charge q the force is $\mathbf{F} = q(\mathbf{v} \times \mathbf{B})$. Therefore the field in the wire due to the movement through $\mathbf{B}$ is $\mathbf{F}/q$ or $\mathbf{v} \times \mathbf{B}$. This is true for any orientation of the field or the wire. The field is of the e.m.f.-producing type.

For the situation of Figure 11.5 $\mathbf{v}$ and $\mathbf{B}$ are at right angles and therefore the field E is upwards and has magnitude vB. E is also the same for all parts of the wire and parallel to l, and therefore $\mathbf{E} \cdot \mathbf{dl} = E \, dl$. Therefore the e.m.f. across the wire is

$$e = \int_0^l E \, dl = vBl \tag{11.5}$$

In the general case where the field E may vary along the wire and $\mathbf{v}$ and $\mathbf{B}$ are at an arbitrary angle the expression for the e.m.f. is

$$e = \int_l \mathbf{E} \cdot \mathbf{dl} = \int_l (\mathbf{v} \times \mathbf{B}) \cdot \mathbf{dl} \tag{11.6}$$

It is possible for both transformer induction and motional e.m.f. to be present together; then both equations (11.4) and (11.6) are needed. The use of motional e.m.f. ideas is best applied to problems when single conductors not forming a loop are moving through a magnetic field. When loops are present Faraday's law is a convenient method of tackling a problem. The following problems illustrate these ideas.

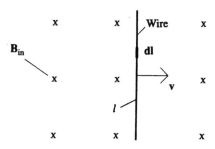

Figure 11.5 A conducting rod moving through a magnetic field

Example 11.4

A straight wire of length $l = 25$ cm is pulled through a magnetic field with a velocity of 0.4 m/s (Figure 11.6). The velocity is at an angle of 60° with respect to B and B has a value of 2 T. What is the e.m.f. generated between the ends of the wire?

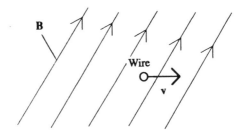

Figure 11.6

The field generated is constant along the wire length and of value $E = v \times B$. This has a magnitude $vB \sin 60°$ and acts along the wire out of the paper.

The magnitude of the e.m.f. between the ends of the wire is given by

$$e = (vB \sin 60°)l = (0.4)(2)(0.866)(0.25) = 173.2 \text{ mV}$$

Example 11.5

A magnetic field with B into the paper extends over a region defined by the length b. A wire loop with sides l and h is pulled through the field region at a speed v (Figure 11.7). The loop is connected to a resistor R. What current flows in the resistor?

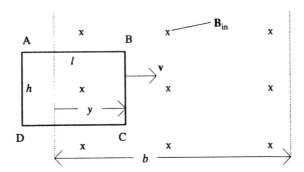

Figure 11.7

Three situations arise:

1. The loop totally immersed in the field.
2. The loop moving across the left field boundary.
3. The loop moving across the right field boundary.

1. In this situation the flux through the loop remains constant. The e.m.f. induced is zero and there is no current.
2. Let the distance from BC to the field edge be y. y is a variable and $dy/dt = v$ since y gets bigger as the loop moves to the right.

   The flux enclosed by the loop is Bhy and the e.m.f. magnitude generated is $d(Bhy)/dt = Bhv$. The current induced in the loop is Bhv/R.

   This current is due to the conductor BC. There is no contribution from AB and CD since there is no e.m.f. generated along the length of these conductors. Conductor DA is not in the field. The direction of the e.m.f. and current is such that the loop resists the movement. The force on a current in a field B is given by $I(\mathbf{l} \times \mathbf{B})$. The current flow must be anticlockwise as this causes a force to act on BC that opposes the movement. The forces due to currents in AB and CD cancel.
3. Let y now be the distance of AD from the field boundary. In this situation the only difference from situation 2 is that y is decreasing as the loop moves and therefore the e.m.f. will be in the opposite direction. The current reverses, and now the force against movement is due to conductor DA.

11.3 Inductance

Consider two coils in proximity (Figure 11.8). A current in coil 1 causes a flux that also cuts coil 2. If the flux is changing then an e.m.f. is induced in coil 2. This is called *mutual induction* and we say there is *mutual*

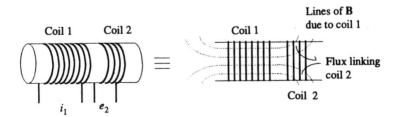

Figure 11.8 Two solenoids in proximity are mutually coupled due to the linking flux

inductance between the two coils. The symbol used for mutual inductance is *M* and the unit is the *henry*.

If a single coil is circulated with a changing current then the flux produced cuts the turns of the single coil and causes a self-induced e.m.f. The phenomenon is called *self-induction* and we say the coil has *self-inductance*. The symbol for self-inductance is *L* and the unit is the *henry*.

Suppose the flux linking each turn of a single coil is Φ, the coil has *n* turns, and a current *i* is flowing. Then according to Faraday's law an e.m.f. is generated across the coil with value $e = -\mathrm{d}(n\Phi)/\mathrm{d}t$. For a given isolated coil the number of flux linkages $n\Phi$ is proportional to the current *i*. If the proportionality constant is denoted by the symbol *L* then

$$n\Phi = Li \tag{11.7}$$

Therefore

$$e = -L\,\mathrm{d}i/\mathrm{d}t \tag{11.8}$$

The self-inductance *L* has units volts seconds/amps or henries.

A comment concerning the minus sign in equation (11.8) is necessary since in Part 2 the equation $v = +L\,\mathrm{d}i/\mathrm{d}t$ is quoted. In Part 2 *L* is assumed to be a passive component in a circuit and the convention is shown in Figure 11.9. In equation (11.8) the e.m.f. indicated is a source e.m.f. and therefore is reversed in sign.

Figure 11.9 Sign conventions for current and voltage for an inductor

In a similar way, for mutual induction, e_2, the e.m.f. in coil 2 due to a changing current i_1 in coil 1, is written as

$$e_2 = -M\,\mathrm{d}i_1/\mathrm{d}t \tag{11.9}$$

The directions and signs are only really important when a specific situation is postulated. In particular, as shown in Part 2, for mutual inductance effects a *dot* notation is used to deal with the sign problem.

11.3.1 Inductance of a long solenoid

Consider a solenoid with area *A*, length *l*, and having *n* turns/m (Figure 11.10). The solenoid carries a constant current *I*.

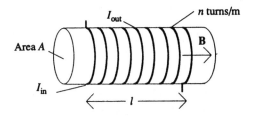

Figure 11.10 A solenoid with *n* turns per metre

For a solenoid with an air core the magnetic flux density inside the solenoid is given by equation (10.13) and is $B = \mu_0 nI$. For a total number of turns N and flux Φ the flux linkages are

$$N\Phi = (nl)(BA) = LI$$

Therefore

$$L = \mu_0 n^2 lA \tag{11.10}$$

This is an important result since the solenoid is the normal method of producing a practical inductance.

11.3.2 Inductance of a coaxial cable

Coaxial cables are much used in telecommunications and electronics. The inductance per metre length of coaxial cables is an important feature.

Consider a cable with the dimensions shown in Figure 11.11. The technique here is to find B at radius r and from this the total flux. Current travels down the centre conductor and back along the outer; therefore a loop surrounds the dielectric and the flux flows through the dielectric as shown.

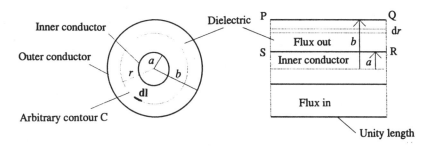

Figure 11.11 A coaxial cable with contour at radius *r*

To find the flux density at radius r

Let I be the current in the conductors; then by Ampère's law

$$\oint_C \mathbf{B} \cdot \mathbf{dl} = \mu_0 I$$

Now **dl** is a short length on contour C and is everywhere parallel to the field **B** which is concentric. B is also constant on contour C. Therefore since only the current in the centre conductor is enclosed by contour C, Ampère's law reduces to

$$B \oint_C dl = \mu_0 I$$

and

$$B = \mu_0 I / 2\pi r \tag{11.11}$$

This is the same expression as that found for the field round a straight wire, which is not surprising!

The total flux in region PQRS which is contained by the loop of current can be found and L introduced by the flux linkage formula.

To find total flux

Consider a small area of width dr at radius r and length unity along the cable. Since dielectrics have a relative permeability of unity the flux through this area is

$$d\Phi = (\mu_0 I / 2\pi r) dr$$

Therefore

$$\Phi = (\mu_0 I / 2\pi) \int_a^b dr/r$$

$$= (\mu_0 I \ln(b/a))/2\pi$$

Now $LI = N\Phi$, and $N = 1$. Therefore

$$L = (\mu_0 \ln(b/a))/2\pi \tag{11.12}$$

The units for L in equation (11.12) are H/m. As expected the result depends on the dimensions of the cable and the permeability of the dielectric. Note the similarity to the result for the capacitance of a coaxial cable given in equation (9.21).

―――― **Example 11.6** ―――――――――――――――――――――――――

The dimensions of a typical coaxial cable are: radius of inner conductor 0.25 mm, and radius of outer conductor 1.6 mm. The dielectric constant of the material between the conductors is 2.2. Find values for the capacitance and the inductance per unit length of the cable.

Two values of importance for coaxial cables are the *characteristic impedance* $Z_0 = (L/C)^{\frac{1}{2}}$, and the *velocity of propagation* $v = 1/(LC)^{\frac{1}{2}}$. Determine values for Z_0 and v and comment on the results.

Substituting in equation (9.21):

capacitance = 66 pF/m

Substituting in equation (11.12):

inductance = 0.37 μH/m

Therefore the characteristic impedance is

$Z_0 = 75\ \Omega$

and the velocity of propagation

$v = 2.02 \times 10^8$ m/s

The approximate impedance of a half-wave dipole antenna is 75 Ω. When this coaxial cable is connected to such an antenna there is an impedance *match*. Therefore there is maximum transfer of power from the cable to the antenna.

v is the speed of electromagnetic waves in the dielectric of the cable and is approximately two-thirds of 3×10^8 m/s, the velocity of electromagnetic waves in free space. A pulse travels down the cable at this very high velocity!

11.4 Energy in the magnetic field

Consider a perfect coil with zero resistance and inductance L. A voltage v is applied to the coil. Then the coil develops a 'back e.m.f.' due to its self-inductance which must equal the applied voltage. If the current flow is i then

$$v = L\ di/dt$$

The power in the coil is

$$vi = Li\ di/dt$$

If energy is W then

$$dW/dt = Li\ di/dt$$

Rearranging

$$dW = Li\,di$$

Integrating

$$W = Li^2/2 \qquad\qquad (11.13)$$

Notice the similarity to other energies such as kinetic $(mv^2/2)$ and capacitor $(CV^2/2)$. For an inductor of 100 mH carrying a current of 0.4 A the energy is 8 mJ. This energy is present in the field surrounding the coil and therefore it is of interest to find the energy density in this region.

11.4.1 Energy density in a magnetic field

Consider a long solenoid where the flux is concentrated inside and the volume of the device is known. A solenoid with dimensions is shown in Figure 11.12. The solenoid has an air core.

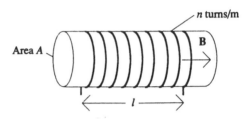

Figure 11.12   A section of a long solenoid of known dimensions with the magnetic field concentrated inside

As proved previously a solenoid of these dimensions carrying a current i has flux density $B = \mu_0 in$ and inductance $L = \mu_0 n^2 lA$. Dividing energy by volume

$$\text{energy density} = Li^2/2Al = B^2/2\mu_0 \qquad\qquad (11.14)$$

The units of energy density are J/m^3.

As expected this is independent of coil dimensions and dependent only on the flux density and the magnetic material. It also has a similar form to the electrical energy density $\varepsilon_0 E^2/2$.

The result of equation (11.14) is true for any magnetic field in free space and is independent of the manner by which the flux is produced.

_____   **Example 11.7** _____

Find the maximum magnetic energy density in the dielectric of a coaxial cable with dimensions $a = 0.42$ mm, $b = 2.5$ mm, and which carries a current of 500 mA. Describe in words how the energy per unit length may be determined.

B in a coaxial cable changes with radius and is a maximum at the inner conductor surface. The dielectric has a relative permeability of unity. At the inner conductor surface where the radius is a, $B = \mu_0 I/2\pi a$. Therefore the maximum energy density is

$$B^2/2\mu_0 = 22.5 \text{ mJ/m}^3$$

It is necessary to find the energy in an infinitesimal area of unit length at an arbitrary radius r and then integrate over the dielectric area. (See problem P11.14 at the end of the chapter.)

_____ **Problems** _____

P11.1 A magnetic flux of value 1250 μWb is passed through a coil of 500 turns. If this flux is reversed in 0.05 s determine the e.m.f. induced in the coil.

P11.2 A coil of 10 turns and area 0.2 m² in air is oriented at right angles to an alternating magnetic field of frequency 5 MHz. The field has a peak value B of magnitude 5×10^{-10} T. Determine the value of the e.m.f. induced in the coil.

P11.3 A current of 5 A flows in the long solenoid that surrounds the small 50 turn coil in Figure P11.1. The current is changed linearly from 5 A to 0 A in 0.04 s.

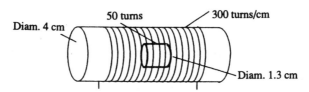

Figure P11.1

Determine the e.m.f. induced in the small coil.

P11.4 A conducting rod of length 10 cm is moved at a velocity of 0.4 m/s at right angles to its length. If the rod moves at right angles to a magnetic field with flux density 0.25 T, determine the e.m.f. induced in the rod. What determines the direction of the induced e.m.f.?

P11.5 A conducting rod is moved at a velocity of 2 m/s parallel to a long straight wire carrying a current of 40 A as shown in Figure P11.2.
    Find the e.m.f. generated in the rod.

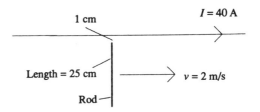

Figure PII.2

P11.6 A rectangular single-turn loop with the dimensions shown in Figure P11.3 is moved at a velocity of 0.4 m/s across a region of magnetic flux with density 0.2 T.

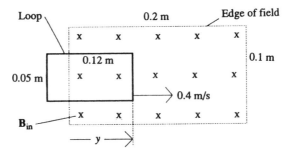

Figure PII.3

Determine the e.m.f. generated in the loop as it moves across the field area. Plot a graph of the e.m.f. against distance y for y values from -0.15 m to 0.35 m.

P11.7 In the arrangement of the previous question a resistance of $10\,\Omega$ is connected across the loop. Find the power generated in the loop as it is moved across the magnetic field.

Show that the forces acting on the loop, due to the current flowing, obey Lenz's law and act against the 'mover'. Also demonstrate that the power used by the 'mover' balances the power generated in the loop.

P11.8 (a) A toroid with 1000 turns is wound on a non-magnetic ring with a radius of 10 cm and a square cross-sectional area of 0.4 cm². Find the inductance of the toroid.

(b) A magnetic circuit is designed with a reluctance of value $10^6\,H^{-1}$. A solenoid with 1200 turns is wound on this magnetic circuit. Determine the inductance of the solenoid.

P11.9 Find an accurate expression for the magnetic flux density in the centre and on the axis of a solenoid by using Biot–Savart's law. Assume the solenoid has a length l, n turns/m, and carries a current I. Show also that this result degenerates to $\mu_0 nI$ when $l \gg R$, where R is the radius of the solenoid.

(Hints: Assume the solenoid turns can be replaced by a current sheet of length L. Assume the definite integral

$$\int_{-b/2}^{b/2} dx/(x^2 + a^2)^{3/2} = 2b/[a^2(4a^2 + b^2)^{1/2}].)$$

P11.10 Consider the solenoid of problem P11.9. Using the result for B found in P11.9 find an expression for the 'true' self-inductance L of the solenoid. Assume B is uniform over the solenoid cross-section. Find a value for L when $l = 10$ cm, $R = 2$ cm, and $n = 1000$. Compare with the value obtained using the simplified formula (equation (11.10)).

P11.11 Find an expression for the self-inductance of a twin-wire transmission line when the distance between the wire centres is D and the radius of the wires is a (Figure P11.4).

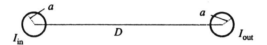

Figure P11.4

Ignore the contribution to the inductance due to the flux in the wires themselves. This is normally termed the *internal inductance*.

(Hint: Find the flux between the lines. Then use equation (11.7) to introduce L.)

P11.12 Determine the magnetic energy per unit length in a straight conductor of radius a carrying a current I. Assume I is distributed uniformly. Using this result show that the 'internal' self-inductance of the wire is given by the expression $\mu_0/8\pi$ H/m. This self-inductance is very small ($0.05\ \mu$H), and is due to the flux in the conductor; it is often neglected when doing inductance calculations on more complex structures (see problem P11.11).

Why do you think the 'internal' inductance is independent of the size of the wire?

(Hint: There is flux in the wire due to the current and this flux links a variable part of the total current.)

P11.13 Find an expression for the mutual inductance between a long straight wire and a square loop of wire with n turns, if the near side of the loop is distance a from the straight wire, and the far side distance b.

Estimate the value of the mutual inductance when $n = 100$, $a = 0.5$ cm and $b = 20$ cm.

P11.14 A coaxial cable has a central conductor of radius a, and an outer braid conductor whose internal radius is b. The cable is carrying a current I.

Derive an expression for the magnetic energy stored per unit length of cable. Using this result, determine an expression for the inductance of the cable per unit length. Find values for the inductance and energy when $b = 3.5$ mm, $a = 1$ mm, and $I = 1$ A.

CHAPTER TWELVE

Maxwell's equations

Maxwell's equations have all been introduced in the previous chapters. This chapter serves to collect the equations and emphasize their importance. However, one equation, namely Ampère's law, still needs further discussion. Maxwell realized that although Ampère's law as stated in equation (10.10) is true for static phenomena, it does not fully describe the time-varying case. Maxwell called the necessary additional term in Ampère's law the *displacement current*; it is easily demonstrated using a simple capacitor circuit.

12.1 Displacement current

Consider the capacitor in Figure 12.1, with a time-varying voltage applied. A time-varying current i flows round the circuit. In the wire connecting the source to the capacitor the current is due to conduction electrons in the wire. The dielectric in the capacitor has no conduction electrons, so what is the nature and source of the current in the capacitor?

Assume the capacitor is of the parallel plate type, with area A, plate separation d, and dielectric constant ε. The charge on C is

$$q = Cv$$

Therefore

$$i = dq/dt = C \, dv/dt$$
$$= (\varepsilon A/d)(dE/dt)(d)$$
$$= \varepsilon A \, dE/dt$$

This is called the *displacement current* after Maxwell. For this simple capacitor circuit the current in the circuit is the same everywhere and so the conduction current in the wire is equal to the displacement current in

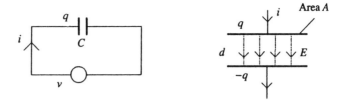

Figure 12.1   Displacement current in a capacitor

the capacitor. The source of the 'current' in the capacitor is the *time-varying electric field*. The current is often written in terms of the electric flux density as $A \, dD/dt$.

The magnetic field round the wire outside the capacitor still exists inside the capacitor owing to the displacement current. A rate of change of electric field can therefore cause a magnetic field and must be part of Ampère's law. A complete statement of Ampère's law, formerly given in equation (10.10), is

$$\oint_C \mathbf{B} \cdot \mathbf{dl} = \mu_0 (I + \varepsilon A \, dE/dt) \tag{12.1}$$

When current density is specified rather than current the equation becomes

$$\oint_C \mathbf{B} \cdot \mathbf{dl} = \mu_0 \int_S (\mathbf{J} + \varepsilon \, d\mathbf{E}/dt) \cdot \mathbf{ds} \tag{12.2}$$

For static phenomena $d/dt = 0$ and equations (12.1) and (12.2) degenerate to the old forms.

_____ **Example 12.1** _____

Determine the peak displacement current density due to an electric field of magnitude 10 kV/m varying sinusoidally at 5 kHz in mica. Mica has a relative permittivity of value 6.

The field can be represented as $10^4 \sin \omega t$. The current density is

$$\varepsilon_r \varepsilon_0 \, dE/dt = 6(8.85.10^{-12})10^4(2\pi 5000)\cos \omega t$$
$$\text{the peak value} = 16.7 \text{ mA/m}^2$$

Note how small the current density is for a large field of 10 000 V/m. Increasing the field frequency will increase the current density.

12.2 Maxwell's four integral equations

The four equations are extracted from sections 9.6, 10.5, 11.1, and 12.1 and are stated in *integral form* in free space as

$$\oint_S \mathbf{E} \cdot \mathbf{ds} = q/\varepsilon_0 \qquad \text{Gauss's law} \qquad (12.3)$$

$$\oint_S \mathbf{B} \cdot \mathbf{ds} = 0 \qquad \text{Gauss's law} \qquad (12.4)$$

$$\oint_C \mathbf{B} \cdot \mathbf{dl} = \mu_0 \int_S (\mathbf{J} + \varepsilon \, d\mathbf{E}/dt) \cdot \mathbf{ds} \quad \text{Ampère's law} \qquad (12.5)$$

$$\oint_C \mathbf{E} \cdot \mathbf{dl} = -d/dt \int_S \mathbf{B} \cdot \mathbf{ds} \qquad \text{Faraday's law} \qquad (12.6)$$

Other relations already defined which may be used with Maxwell's equations are

$$\mathbf{D} = \varepsilon \mathbf{E} \quad \mathbf{B} = \mu \mathbf{H} \quad \mathbf{J} = \sigma \mathbf{E}$$

where ε, μ and σ are dependent on the materials involved, and

$$\mathbf{F} = q(\mathbf{E} + \mathbf{v} \times \mathbf{B})$$

which effectively defines $\mathbf{E}$ and $\mathbf{B}$ in terms of force and charge.

The scope of these equations is enormous; they cover all non-relativistic phenomena associated with the electromagnetic field. This includes electrical rotating machinery, electromagnetic levitation, particle acceleration, radio, radar, television, microwaves, and optics.

The integral equations are a good starting point for the study of *electromagnetic waves* at second-year degree level. This in turn gives an excellent grounding for more advanced work in final-year specialist courses on topics such as *microwaves* and *optical fibres*. These latter topics are of course of great importance in communications.

CHAPTER THIRTEEN

Relationship between circuits and fields

Circuits deal with *idealized circuit components* with values of resistance *R*, self-inductance *L*, mutual inductance *M*, and capacitance *C*. It is common to measure the current *I* through the components and the voltage *V* across them.

Fields deal with the *field quantities E and B* and their magnitude and direction as a function of position. Components are treated as three-dimensional objects for which field quantities can be derived.

13.1 Circuits

The assumption that it is possible to make a resistor that has resistance only, an inductor that has inductance only, and a capacitor that has capacitance only, is a good practical approximation at '*low*' frequencies. By specifying a device by the idealized value, there is no need to consider its practical construction and three-dimensional problems are eliminated.

Connections are made by copper wire that is assumed to have zero resistance, again a good practical approximation at low frequencies. The wires guide the current to each component and so it is not necessary to worry about the spatial direction of current flow.

For these reasons circuits do not need a vector description. Therefore whereas in fields a vector form of Ohm's law $\mathbf{J} = \sigma\mathbf{E}$ is needed, in circuits a scalar law $V = IR$ suffices.

A full rigorous analysis of a component would have to utilize all the three qualities of resistance, inductance, and capacitance. The circuit impedances of ideal passive components are R, $j\omega L$, and $1/j\omega C$; even if L and C are very small, large values of ω can cause the impedances to have a large effect on the circuit. Nevertheless the real component can still be *modelled* using idealized components.

217

The effects of frequency on components are quantified in the example in the following section.

13.1.1 A practical inductor

Consider a simple air-cored inductor in the form of a long solenoid. Suppose the inductor has the dimensions shown in Figure 13.1.

The inductance of the coil is given by $L = \mu_0 n^2 lA$ where $A = \pi r^2$. For the values shown L has a value of 1.97 mH.

The resistance R of the coil is given by $h/\sigma a$. The total length of wire on the coil is $h = 2\pi r(1000) = 10\pi$ m, the cross-sectional area of the wire is $a = \pi \cdot 10^{-8}$ m^2, and the conductivity of copper σ is 5.7×10^7 S/m. Therefore the resistance R has the value 17.5 Ω.

The coil will also have capacitance C due to the proximity of individual turns. The calculation for C is complicated; the model should now be a distributed-parameter one. Let us, however, assume that a crude approximation to the capacitance effects can be modelled by a single capacitor across the coil. A practical estimate for this coil is $C = 50$ pF.

Overall the inductor has the equivalent circuit of Figure 13.2. The impedance Z of the components as frequency is varied is shown in Figure 13.3. Note that the resistance of the inductor is comparable with the inductive impedance in the low frequency kilohertz range. The capacitance is significant at high frequencies and is comparable with R in the gigahertz region. The inductance and capacitor are comparable in the megahertz range.

It is clear from this graph that the behaviour of the so-called inductor depends crucially on the frequency of operation. In particular at high frequencies the effect of C is very pronounced, and a different approach is needed to design components and develop satisfactory circuit behaviour at high frequencies.

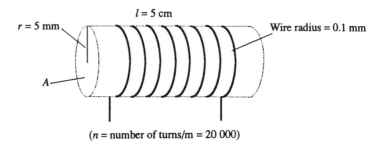

$(n = \text{number of turns/m} = 20\ 000)$

Figure 13.1   A solenoid with specific dimensions

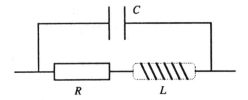

Figure 13.2 A model for a practical inductor using idealized circuit elements

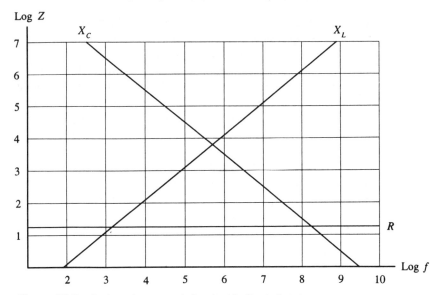

Figure 13.3 An impedance graph for the idealized elements

13.2 Fields

The level of frequency at which circuit concepts fail needs clarifying. Velocity is equal to wavelength multiplied by frequency, and in vacuum the velocity of electromagnetic waves is 3×10^8 m/s. At low frequencies the wavelength is very large compared with the circuit size; at a frequency of 50 Hz, $\lambda = 6 \times 10^6$ m. As a rough rule of thumb problems arise when the wavelength starts to become *of the same order* as the circuit size. The difficulties associated with high frequency work can be appreciated by noting that at microwave frequencies, say 10 GHz, the wavelength is 3 cm. At such frequencies it is necessary to use field concepts for component and system design.

Consider the design of a tuned circuit at 10 GHz. A classical *LC* network would require values of the order of $L = 0.01 \, \mu$H and $C = 0.03$ pF. Individual conventional circuit components with such small values cannot be made accurately; the stray values of C and L to other components are unknown and likely to be bigger than the values of the component itself. Instead a tuned cavity designed using field quantities such as $\mathbf{E}$ and $\mathbf{B}$ would be appropriate. Such designs may be found in books on microwaves.

13.3 An *LCR* circuit with a field description

A conventional circuit is analyzed using field methods and the results shown to be equivalent to those using circuit methods.

Let us consider a series circuit utilizing the three *ideal* passive components R, L, and C (Figure 13.4). Assume a current i flows round the circuit. The current is directed by the connecting wires through all components and therefore *Kirchhoff's current law* is automatically satisfied. *Kirchhoff's voltage law* can be checked by finding the voltage drops around the circuit in field terms and simplifying to the circuit expressions. The field expression for the voltage drop is $\int \mathbf{E} \cdot \mathbf{dl}$.

The wires are assumed to have a conductivity σ which is infinite. Then the field expression $J = \sigma E$ gives zero E for finite J. This implies zero voltage drop along the connecting wires. This applies also to the wire with which the inductor is wound.

For the complete circuit

$$\oint_C \mathbf{E} \cdot \mathbf{dl} = 0$$

Adding up all the voltage drops around the circuit we have

$$\int_{12} \mathbf{E} \cdot \mathbf{dl} + \int_{23} \mathbf{E} \cdot \mathbf{dl} + \int_{34} \mathbf{E} \cdot \mathbf{dl} + \int_{41} \mathbf{E} \cdot \mathbf{dl} = 0$$

Let us consider the terms separately.

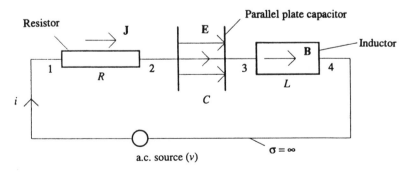

Figure 13.4 A simple *LCR* series circuit

The source

$$v_4 - v_1 = \int_{41} \mathbf{E} \cdot \mathbf{dl}$$

$$= -v \quad \text{(the source e.m.f.)}$$

The resistor

Assume a resistor of length l, area a, and conductivity σ; then

$$v_1 - v_2 = \int_{12} \mathbf{E} \cdot \mathbf{dl}$$

$$= (1/\sigma) \int \mathbf{J} \cdot \mathbf{dl}$$

$$= Jl/\sigma$$
$$= i(l/\sigma a)$$
$$= iR$$

This is the standard circuit expression.

The capacitor

Let the capacitor have plate area s, plate separation d, and charge q. Using Maxwell's first equation, the field E is given by $\int \mathbf{E} \cdot \mathbf{ds} = q/\varepsilon$ which simplifies to $E = q/\varepsilon s$. Then the voltage drop is

$$v_2 - v_3 = \int_{23} \mathbf{E} \cdot \mathbf{dl}$$

$$= Ed$$
$$= q(d/\varepsilon s)$$
$$= q/C$$

This is the standard circuit expression.

The inductor

Let us assume the inductor is a long solenoid with area A, length h, and with n turns/m. Such an inductor has a magnetic flux density inside given by $B = \mu_0 ni$ and all turns are cut by it. The self-inductance L is given by $\mu_0 Ahn^2$. For this simple case $\int_s \mathbf{B} \cdot \mathbf{ds} = BA$ and therefore using Maxwell's fourth equation

$$v_3 - v_4 = -\int_{34} \mathbf{E} \cdot \mathbf{dl}$$

$$= (nh)d/dt \int_s \mathbf{B} \cdot \mathbf{ds}$$

$$= A(nh)\mathrm{d}B/\mathrm{d}t$$
$$= (An^2h\mu_0)\mathrm{d}i/\mathrm{d}t$$
$$= L\,\mathrm{d}i/\mathrm{d}t$$

This is the standard circuit expression.

The overall circuit

Noting that $q = \int i\,\mathrm{d}t$, the complete circuit equation is

$$L\mathrm{d}i/\mathrm{d}t + Ri + \int i\mathrm{d}t/C = v$$

The field descriptions lead to the basic circuit equation for a series circuit consisting of the three ideal passive elements.

Components: modelling, preferred values, and tolerances

It has been emphasized that circuit analysis is about the analysis of a circuit diagram. The results of the analysis should be precise and accurate. How well the analysis predicts the actual performance of the real circuit depends upon the accuracy with which the real circuit is modelled by the ideal components of the circuit analysis diagram. Some physical situations can only be analysed using a fields approach and this is discussed in Chapter 13. Some components have parameters which are uniformly distributed throughout the component; for example, the resistance of a transmission cable is distributed uniformly along its length. The concept of length does not arise in the ideal resistance of circuit analysis. It is sometimes referred to as a *lumped-constant* model, the effects of the parameter resistance being regarded as being lumped together at one point. It is possible to approximate a model of a component with distributed parameters by representing a small part of the component by an ideal circuit element, or elements, and to model the component by the combination of a large number of such elements. A good example of this is the circuits approach to the analysis of a transmission cable. It is recognized that a cable will add some resistance and some inductance in series with a signal path and will add some capacitance across or in parallel with the path. A small length l of the cable may thus be modelled by the circuit shown in Figure A1.1.

A practical length of cable is represented by many such circuits in cascade. The smaller the length l and the more circuits used the greater the accuracy of the analysis. In any problem involving distributed parameters a judgement is needed as to the number of repeated circuits required to give a usefully accurate result. A small number can give a manageable analysis and a large number can sometimes be handled algebraically by deducing a limiting value of some parameter of interest as l is allowed to approach zero. In other cases it may be concluded that a fields approach is a more appropriate analysis.

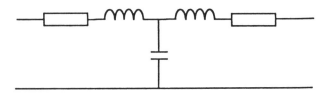

Figure AI.I   Circuit model for a small length of transmission cable

Another aspect of modelling concerns the practical components: the resistor, the capacitor, and the inductor. For many commonly occurring circuits a resistor may be modelled by an ideal resistance. However, a resistance may have some inductive properties; in a wire-wound resistor this can be significant, and might have some capacitance between its ends. The significant factor in all components is the relative effect of the undesired property compared with that of the major property. At very high frequencies the series inductance and shunt capacitance may need representing in the model of the resistor. The same sort of considerations apply to a capacitor. For most purposes it can be represented by an ideal capacitance but in particular cases detailed data may be needed, obtained by measurement or from the component manufacturer, in order to produce a more accurate model. A good example is that of an electrolytic capacitor where some series resistance and possibly inductance may be required to give an accurate model. The inductor is in principle treated in the same way but it is impossible to make an ideal inductance as the resistance of the wire is not usually negligible and core losses need to be accounted for if the inductor is not air-cored. A more detailed treatment of this is given in Chapter 13. Transformers and active devices give rise to similar considerations with many levels of model being used. They will be regarded as being beyond the scope of this book.

Another practical consideration arising is that of the tolerance on the value of the component achieved in manufacture. In introducing analysis methods precise values are always assumed for the circuit elements. However, and this is important when circuit design is considered, components will normally have a nominal value and a specified tolerance on this within which the actual value will lie. A parameter of interest in the design of a circuit is the sensitivity of a particular property of the circuit to changes in component value. In general for a given requirement the better design will be less sensitive to changes in component value. Algebraic analysis can become quite difficult; inequalities need to be solved rather than equations. Graphical and phasor diagram solutions can be very helpful but the number-crunching power of the computer is becoming the best solution to this problem. A final practical point to raise, again affecting design more than analysis, is the question of preferred values. Taking, for example, resistors it is found that only certain preferred values are available. For resistors with a tolerance of 10% the preferred values are 1.0, 1.2, 1.5, 1.8, 2.2, 2.7, 3.3, 3.9, 4.7, 5.6, 6.8, and 8.2, all multiplied by 10^n where $n = 0$, 1, 2, 3, etc. The details for other tolerances and for capacitor values can be obtained from manufacturers' or suppliers' catalogues.

Time and frequency domain analysis

These terms are introduced in Chapter 8. The distinction is sometimes a matter of initial confusion to the student. In the context of introductory circuit analysis it can be stated quite simply that the relationships between voltage and current for the circuit elements are defined using functions of time, e.g. $v = L \, di/dt$.

The sinusoid is a particular time function that has such wide application that specific techniques have been developed, e.g. 'j' notation. It must be realized that these techniques are only of use for analysis of the steady state condition of circuits with a sinusoidal excitation. If non-sinusoidal waveforms are involved or the transient response of a circuit to a sinusoidal excitation is required then the fundamental time-dependent expressions must be used. Ohm's law and Kirchhoff's laws will always be valid.

The terms time domain and frequency domain also arise in systems work and in electronics. A signal may be defined as a function of time or, provided it is periodic, as a number of sinusoidal components of particular frequency, amplitude, and phase.

In a practical measurement context the oscilloscope displays the amplitude of a signal against time, i.e. a time domain display. The spectrum analyser displays a signal as amplitude against frequency, a frequency domain display.

For all steady state sinusoidal problems encountered at the level of this book the reader is advised to use phasor diagrams or 'j' notation as these are likely to produce the most straightforward solutions.

Computer-aided analysis and simulation

There are many commercially available computer programs which attempt to provide an environment in which the user can describe a circuit and then ask that a particular analysis is performed. Three of these are mentioned as examples with differing merits.

PSpice, (MicroSim Corporation, 20 Fairbanks, Irvine, California, USA)

This is a version of the original spice program which has been adapted for use on a personal computer. An evaluation version is available which restricts the size of the circuit to be analysed but which may be freely copied. Circuits are normally described in spice programs using a word processor or text editor. This means that the circuit must be drawn on paper and numbers allocated to all the circuit nodes. The circuit description is then a list of statements giving at least a label and a value for each component and also the node numbers between which it is connected. More recently schematic capture programs have been introduced which allow the circuit to be drawn on screen using a library of component icons or shapes.

MICRO-CAP (Spectrum Software, Sunnyvale, California, USA)

This program uses spice algorithms in its analysis but has a good circuit-drawing facility and an easily used, menu-driven, facility for displaying the results. A student version is available which, with an instruction manual, costs a little more than an average textbook. This limits the size of the circuit to be analysed but otherwise includes most features of the professional version that would be of interest to the student.

227

ELECTRONICS WORKBENCH (Interactive Image Technologies, Toronto, Canada)

This program has a different philosophy in that it attempts to simulate an electrical workbench with on-screen icons for components and instruments. It has a very good circuit-drawing facility and has some useful features not possible in the other programs mentioned. However, it presents its results on the face of the simulated measuring instrument.

Many other such programs exist. Most of them use spice algorithms and many enable spice netlists, i.e. spice circuit descriptions, to be produced.

The reader is encouraged to use such programs as they provide a valuable learning aid. Once the circuit has been entered it is easily changed; the 'what if' question can be posed and the effect of variation of component values can be readily explored. However, it must be emphasized that such programs are not an alternative to thought or to algebraic manipulation. The user must ultimately understand what he or she is doing. The computer will give wrong answers with great speed and precision if wrong questions are posed. A single correct numeric answer does not provide the insight to circuit operation that can be obtained from algebraic analysis. Nevertheless as long as these dangers are realized exploration of a circuit using such programs can significantly enhance understanding of its operation. No specific examples are included here. There are several books devoted to the use of spice, such as those by Keown (1991), Thorpe (1992), Tuinenga (1992), and Vladimirescu (1994). Also many of the larger circuit analysis texts include worked examples: Bogart (1992), Sander (1992), and Nilsson (1993). The work of section 7.5 on resonance is a good example of a topic that can be explored and verified very easily, both in principle and numerically, using these programs.

Further vectors

This appendix supplements the information given in Chapter 2. The vector definitions given there should be known before reading this appendix.

A4.1 Coordinate systems

A coordinate system is usually chosen to reflect the symmetry of the problem being dealt with. There are three main coordinate systems: the *Cartesian coordinate system*, the *spherical coordinate system*, and the *cylindrical coordinate system*. The Cartesian system only is used in this text. Discussions on all three sets of coordinate systems may be found in more advanced texts on electromagnetism.

The Cartesian system is shown in Figure A4.1. In this figure the axes are all at right angles, and by convention conform to the rule that a rotation from x to y should cause a right-handed screw to progress along the z-axis. This is called a *right-handed Cartesian coordinate system*.

Figure A4.2 shows this coordinate system with a vector OA drawn from the origin to an arbitrary point in space A. A perpendicular from A meets the xy plane at point B. The position of this vector with respect to axes xyz is specified in terms of *unit* vectors.

A4.2 Unit vectors

A unit vector is a vector specifying a particular direction, and with unity magnitude. For the Cartesian system the unit vectors are denoted by the symbols a_x, a_y, and a_z. In older books the unit vectors were specified as i, j, and k. Therefore a vector of magnitude 12 in the y direction would be specified as $12a_y$ or $12j$.

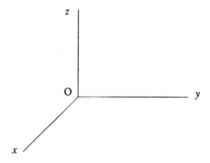

Figure A4.1 Cartesian axes

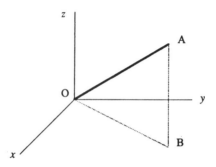

Figure A4.2 Vector OA with perpendicular AB onto the *xy* plane

If the vector OA of Figure A4.2 has magnitude F then this fact can be given in terms of the three magnitudes F_x, F_y, and F_z to reach point A in directions x, y, and z as shown in Figure A4.3.

Then the vector may be written as

$$\mathbf{F} = \mathbf{a}_x F_x + \mathbf{a}_y F_y + \mathbf{a}_z F_z \tag{A4.1}$$

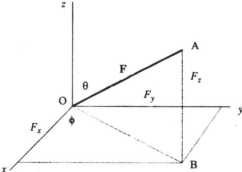

Figure A4.3 **F** with *xyz* components

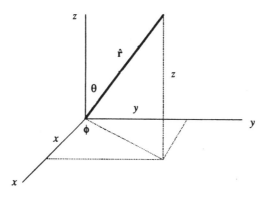

Figure A4.4   r with *xyz* components

In this expression it is important to recognize that the symbols F_x, F_y, and F_z are scalar values specifying magnitudes only and are therefore in normal type. The vectors **F** and **a** on the other hand are printed in bold type.

From Figure A4.3 it is clear that the magnitude of the vector **F** is given by

$$|F| = (F_x^2 + F_y^2 + F_z^2)^{\frac{1}{2}} \tag{A4.2}$$

A unit vector along an arbitrary direction in space is denoted by the symbol $\hat{r}$. Unity magnitude is denoted by the carot symbol (^). Therefore $\mathbf{F} = 10\hat{r}$ means that the vector **F** is of magnitude 10 and acts along the direction specified by $\hat{r}$.

The unit vector $\hat{r}$ is specified in terms of the Cartesian unit vectors as in Figure A4.4. Suppose the scalar values defining $\hat{r}$ along the three axes are x, y, and z; then

$$\hat{r} = \mathbf{a}_x x + \mathbf{a}_y y + \mathbf{a}_z z \tag{A4.3}$$

where

$$(x^2 + y^2 + z^2)^{\frac{1}{2}} = 1 \tag{A4.4}$$

Equation (A4.4) ensures that $\hat{r}$ has unity magnitude irrespective of its direction in space.

_____ **Example A4.1** _____

(a)   What is the magnitude of the vector $\mathbf{F} = \mathbf{a}_x 5 + \mathbf{a}_y 12 + \mathbf{a}_z 9$?
(b)   If the z component is made zero find the new magnitude and make a sketch detailing the position of the new vector with respect to the axes.

(a) $|F| = (25 + 144 + 81)^{\frac{1}{2}} = 15.81$

(b) $|F| = (25 + 144)^{\frac{1}{2}} = 13$

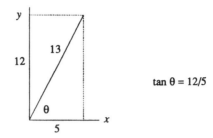

$\tan \theta = 12/5$

Figure A4.5

In the sketch of Figure A4.5 the new vector must be in the xy plane only, with magnitude 13, at an angle $\theta = 67.3°$ to the x-axis.

A4.3 Vector algebra

Vectors may be added and subtracted but unlike scalars the directions must be considered. For the vector description as detailed above addition and subtraction are easy since it is only necessary to work on the three axes independently.

'Multiplication' of vectors is special because of the varying directions and will be dealt with in the next section.

_____ **Example A4.2** _____

Consider the two vectors

$$A = a_x 4 + a_y 5 + a_z 8 \quad \text{and} \quad B = a_x 3 + a_y 5 + a_z 2$$

It is desired to add and subtract these vectors. This is done in the Cartesian system by dealing with the axes independently:

$$A + B = a_x(4 + 3) + a_y(5 + 5) + a_z(8 + 2) = a_x 7 + a_y 10 + a_z 10$$

$$A - B = a_x(4 - 3) + a_y(5 - 5) + a_z(8 - 2) = a_x 1 + a_z 6$$

Clearly in the second case the resultant vector is in the xz plane.

A4.4 Scalar and vector products

These functions take the place of 'normal' scalar multiplication. The direction of the two vectors being 'multiplied' has an effect on the result.

A4.4.1 The scalar product using unit vectors

Remember that $\mathbf{A} \cdot \mathbf{B}$ is a scalar quantity and has magnitude $AB \cos \theta$.
A and B may be represented in unit vector form as

$$\mathbf{A} = \mathbf{a}_x A_x + \mathbf{a}_y A_y + \mathbf{a}_z A_z \quad \text{and} \quad \mathbf{B} = \mathbf{a}_x B_x + \mathbf{a}_y B_y + \mathbf{a}_z B_z$$

Then

$$\mathbf{A} \cdot \mathbf{B} = (\mathbf{a}_x A_x + \mathbf{a}_y A_y + \mathbf{a}_z A_z) \cdot (\mathbf{a}_x B_x + \mathbf{a}_y B_y + \mathbf{a}_z B_z)$$
$$= A_x B_x + A_y B_y + A_z B_z$$

since

$$\mathbf{a}_x \cdot \mathbf{a}_x = \mathbf{a}_y \cdot \mathbf{a}_y = \mathbf{a}_z \cdot \mathbf{a}_z = 1 \quad (\cos 0° = 1)$$

and

$$\mathbf{a}_x \cdot \mathbf{a}_y = \mathbf{a}_z \cdot \mathbf{a}_x = \mathbf{a}_y \cdot \mathbf{a}_z = 0 \quad (\cos 90° = 0)$$

———— **Example A4.3** ————————————————————————

Find $\mathbf{A} \cdot \mathbf{B}$ for $\mathbf{A} = 2\mathbf{a}_x + 5\mathbf{a}_y + 8\mathbf{a}_z$, and $\mathbf{B} = 4\mathbf{a}_x + 6\mathbf{a}_y + 10\mathbf{a}_z$.

Performing the operation $\mathbf{A} \cdot \mathbf{B}$ gives $8 + 30 + 80 = 118$.

A4.4.2 The vector product using unit vectors

Remember that $\mathbf{F} = \mathbf{A} \times \mathbf{B}$ is a vector with magnitude $AB \sin \theta$. The direction is perpendicular to both A and B such that a right-handed screw motion from A to B advances along F.
A and B may be represented in unit vector form as

$$\mathbf{A} = \mathbf{a}_x A_x + \mathbf{a}_y A_y + \mathbf{a}_z A_z \quad \text{and} \quad \mathbf{B} = \mathbf{a}_x B_x + \mathbf{a}_y B_y + \mathbf{a}_z B_z$$

Then

$$\mathbf{A} \times \mathbf{B} = (\mathbf{a}_x A_x + \mathbf{a}_y A_y + \mathbf{a}_z A_z) \times (\mathbf{a}_x B_x + \mathbf{a}_y B_y + \mathbf{a}_z B_z)$$
$$= \mathbf{a}_x (A_y B_z - A_z B_y) + \mathbf{a}_y (A_z B_x - A_x B_z) + \mathbf{a}_z (A_x B_y - A_y B_x)$$

since

$$\mathbf{a}_x \times \mathbf{a}_x = \mathbf{a}_y \times \mathbf{a}_y = \mathbf{a}_z \times \mathbf{a}_z = 0 \quad (\sin 0° = 0)$$

and

$$\mathbf{a}_x \times \mathbf{a}_y = \mathbf{a}_z \quad \mathbf{a}_y \times \mathbf{a}_z = \mathbf{a}_x \quad \mathbf{a}_z \times \mathbf{a}_x = \mathbf{a}_y \quad (\sin 90° = 1)$$

and

$$\mathbf{a}_y \times \mathbf{a}_x = -\mathbf{a}_z \quad \mathbf{a}_z \times \mathbf{a}_y = -\mathbf{a}_x \quad \mathbf{a}_x \times \mathbf{a}_z = -\mathbf{a}_y \quad (\sin 90° = 1)$$

The minus signs arise from the right-handed screw rule.

This result may also be written in determinant form which is easier to remember:

$$\mathbf{A} \times \mathbf{B} = \begin{vmatrix} \mathbf{a}_x & \mathbf{a}_y & \mathbf{a}_z \\ A_x & A_y & A_z \\ B_x & B_y & B_z \end{vmatrix}$$

_____ **Example A4.4** _____

Find the vector product $\mathbf{A} \times \mathbf{B}$ for $\mathbf{A} = 2\mathbf{a}_x + 5\mathbf{a}_y + 8\mathbf{a}_z$ and $\mathbf{B} = 4\mathbf{a}_x + 6\mathbf{a}_y + 10\mathbf{a}_z$.

Give the magnitude and angles associated with the resultant vector.

Performing the above operation we have

$$\mathbf{A} \times \mathbf{B} = \mathbf{a}_x(50-48) + \mathbf{a}_y(32-20) + \mathbf{a}_z(12-20)$$
$$= \mathbf{a}_x 2 + \mathbf{a}_y 12 - \mathbf{a}_z 8$$

The magnitude is

$$(4 + 144 + 64)^{\frac{1}{2}} = 14.56$$

Using the standard angles as shown in Figure A4.3 and also section A4.5:

$$\phi = \tan^{-1} 12/2 = 80.5°$$
$$\theta = \tan^{-1}[(4 + 144)^{\frac{1}{2}}/-8] = -56.7°$$

Sketch this to confirm the position of the resultant vector.

_____

A4.5 Unit vectors and trigonometric representation

A trigonometric representation can be found from the unit vectors. This may be done using the information of Figure A4.3 and equation (A4.1). It is very useful to sketch the results. Consider the vector $\mathbf{F} = 3\mathbf{a}_x + 12\mathbf{a}_y + 4\mathbf{a}_z$ in Figure A4.6.

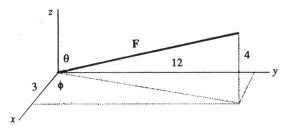

Figure A4.6 A vector **F** with specified component values

The magnitude is

$$|F| = (9 + 144 + 16)^{\frac{1}{2}} = 13$$

This vector makes angles of value

$$\phi = \tan^{-1}(12/3) \quad = 76°$$
$$\theta = \tan^{-1}(12.37/4) = 72°$$

A general example follows that checks unit vector methods against trigonometric methods.

----- **Example A4.5** --

Consider two vectors $A = a_x 4 + a_y 3$ and $B = a_x 5 + a_y 12$ which are in the xy plane only to allow easy sketching.

(a) Find the sum of the two vectors and check the answer using a graph.
(b) Find the scalar and vector products and check the values.

(a) $\quad A + B = a_x(4 + 5) + a_y(3 + 12) = a_x 9 + a_y 15$

A has magnitude

$$(16 + 9)^{\frac{1}{2}} = 5$$

at angle $\tan^{-1} 3/4$ to the x-axis, i.e. 36.9° (see ON in Figure A4.7).
B has magnitude

$$(25 + 144)^{\frac{1}{2}} = 13$$

at angle $\tan^{-1} 12/5$ to the x-axis, i.e. 67.4° (see OM).
The resultant magnitude is

$$(81 + 225)^{\frac{1}{2}} = 17.5$$

at angle $\tan^{-1} 15/9$ to the x-axis, i.e. 59° (see OP).

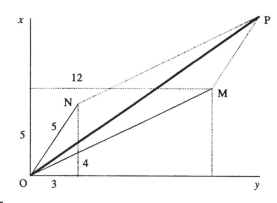

Figure A4.7

(b) $\mathbf{A} \cdot \mathbf{B} = (\mathbf{a}_x 4 + \mathbf{a}_y 3) \cdot (\mathbf{a}_x 5 + \mathbf{a}_y 12)$
$= (\mathbf{a}_x \cdot \mathbf{a}_x)20 + (\mathbf{a}_y \cdot \mathbf{a}_y)36$
$= 56$

since

$$\mathbf{a}_x \cdot \mathbf{a}_y = \mathbf{a}_y \cdot \mathbf{a}_x = 0$$

The magnitude is

$$AB \cos(67.4° - 36.9°) = (5)(13)\cos(30.5°) = 56$$

$\mathbf{A} \times \mathbf{B} = (\mathbf{a}_x 4 + \mathbf{a}_y 3) \times (\mathbf{a}_x 5 + \mathbf{a}_y 12)$
$= (\mathbf{a}_x \times \mathbf{a}_x)20 + (\mathbf{a}_x \times \mathbf{a}_y)48 + (\mathbf{a}_y \times \mathbf{a}_x)15 + (\mathbf{a}_y \times \mathbf{a}_y)36$
$= \mathbf{a}_z 48 - \mathbf{a}_z 15$
$= \mathbf{a}_z 33$

since

$$\mathbf{a}_x \times \mathbf{a}_x = \mathbf{a}_y \times \mathbf{a}_y = \mathbf{a}_z \times \mathbf{a}_z = 0$$

The magnitude is

$$AB \sin(67.4° - 36.9°) = (5)(13)\sin(30.5°)$$
$$= 33$$

and the direction is along the z-axis by the screw rule.

Polarization and magnetization

When a dielectric material is placed in an electric field, dipoles are produced by induction and tend to orient themselves along the field. The dielectric becomes *polarized*.

Similarly when a ferromagnetic material is placed in a magnetic field the magnetic dipoles tend to line up with the field (Figure A5.1) and the material becomes *magnetized*. In both cases the result is that the flux density is *increased* for a given applied field strength. This is important from a practical viewpoint in that improved devices may be made utilizing dielectric and ferromagnetic materials.

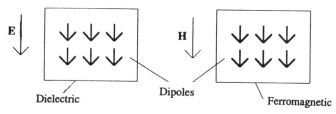

Figure A5.1   Dipoles lining up in the presence of fields

A5.1 Polarization

When discussing polarization effects it is necessary to utilize the electric field vector **E** and the electric flux density vector **D**, together with a new vector **P** to describe the polarization directly. From equation (9.13) we know that $\mathbf{D} = \varepsilon\mathbf{E}$.

Consider two capacitors with the same dimensions, with and without a dielectric material. Let the same *free* charge q be present on both capacitors. The polarization of the dielectric causes a *bound* charge q' as shown in Figure A5.2; q' is bound because it is part of and cannot be

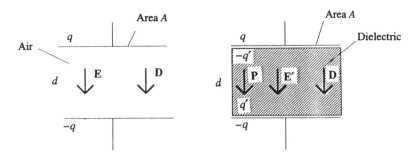

Figure A5.2 Field vectors in a capacitor in air and in a dielectric

removed from the dielectric material. The negative bound charge q' is induced by the positive free charge q attracting the negative ends of the bound dipoles. **E** is due to the *free* charge q. **E'** is due to the *free and bound* charge $(q - q')$ and is therefore less than **E**.

For both diagrams of Figure A5.2 the electric flux density **D** is the same since **D** is due to the *free* charge q only. For the first diagram $\mathbf{D} = \varepsilon_0 \mathbf{E}$. For the second diagram although **D** is the same **E'** is smaller. The difference is made up by the *polarization* vector **P**. The polarization is the effect of the bound charges q', and acts from the negative to the positive bound charge. Therefore for the second diagram we may write $\mathbf{D} = \varepsilon_0 \mathbf{E'} + \mathbf{P}$, and the same **D** is achieved for a smaller value of electric field strength.

This detailed view of the polarization is not usually needed, and is hidden by defining a dielectric constant ε so that for the dielectric case

$$\mathbf{D} = \varepsilon_0 \mathbf{E'} + \mathbf{P} = \varepsilon \mathbf{E'} \tag{A5.1}$$

The effects of polarization are thus included by using the new dielectric constant ε. ε will be larger than ε_0 and the ratio $\varepsilon/\varepsilon_0$ is called the relative permittivity ε_r. As defined in Chapter 9, $\varepsilon = \varepsilon_r \varepsilon_0$.

A5.2 Magnetization

Magnetization, the effect of the internal magnetic dipoles, is given the symbol **M**. For a non-magnetic material there are no internal magnetic dipoles and therefore the magnetization $M = 0$. The flux density is then given by the expression $\mathbf{B} = \mu_0 \mathbf{H}$ as shown in Chapter 10.

In a ferromagnetic region the presence of the field H causes the internal magnetic dipoles to line up in the direction of the applied field. This causes a magnetization M that enhances the flux density for a given field (Figure A5.3).

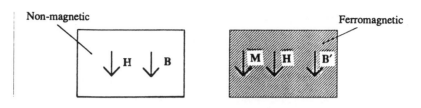

Figure A5.3 Field vectors in magnetic and non-magnetic materials

The flux density in the magnetic material therefore has a greater value that can be written as $\mathbf{B'} = \mu_0\mathbf{H} + \mathbf{M}$. As in the electrical case the detailed view is not often needed and is hidden by defining a new permeability μ so that for the ferromagnetic case

$$\mathbf{B'} = \mu_0\mathbf{H} + \mathbf{M} = \mu\mathbf{H} \tag{A5.2}$$

The effects of magnetization are included by using the new permeability constant μ. μ will be larger than μ_0 and the ratio μ/μ_0 is called the relative permeability μ_r. As defined in Chapter 10, $\mu = \mu_r\mu_0$. μ_r has a value near to unity unless the material is ferromagnetic.

A5.3 Hysteresis

When a ferromagnetic specimen is taken through a cycle of magnetic field strength changes the flux density produced lags the magnetic field strength. This is called *hysteresis* and is shown in Figure A5.4.

The shape of this curve may be explained using the idea of magnetic domains. These are relatively large volumes of the specimen within which dipoles are aligned. The domains change shape and size and orient themselves in the presence of a magnetizing field.

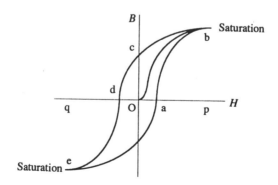

Figure A5.4 A curve of *B* against *H* showing hysteresis

O represents a completely demagnetized initial condition for the ferromagnetic specimen. As *H* is increased along Op the specimen follows line Ob flattening as the magnetic domains become fully aligned with the applied field. Ob is called the *initial magnetization curve*.

When *H* is reduced the domain changes lag those during magnetization and the curve follows bc. Overall as the magnetic field strength follows the path OpOqOp the specimen follows the full hysteresis path Obcdeab.

At c the specimen has residual magnetism without any applied excitation; Oc is called the *remanence*. At d a negative field has been necessary to demagnetize the specimen; Od is called the *coercive force*.

The area traced out by the *hysteresis loop* represents the work done to move the domains in the magnetic material.

The shape of the hysteresis loop can vary very widely according to the material. *Soft* magnetic materials, which are easily magnetized and make poor permanent magnets, have thin loops. *Hard* magnetic materials, which are difficult to magnetize and make good permanent magnets, have fat loops.

Electric and magnetic materials

Any design work requires a knowledge of the properties of the materials used. The basic features of common electromagnetic materials are presented here. All values in the following tables are at standard temperature, pressure, and humidity.

A6.1 Electrical materials

For electromagnetic work the materials of interest are *conductors* and *insulators*. The main feature of the conductor is its *conductivity*. Values are presented in Table A6.1. The insulator is often called *a dielectric*. For dielectrics the qualities of interest are *dielectric constant* and *breakdown field* or *dielectric strength*. Values are given in Table A6.2.

Table A6.1

Material	σ (S/m)
Brass	1.1×10^7
Aluminium	0.5×10^7
Gold	4.1×10^7
Copper	5.7×10^7
Silver	6.1×10^7
Water(d)*	2.0×10^{-4}

Table A6.2

Material	ε_r	Strength (V/m)
Air	1†	3×10^6
Bakelite	4.9	25×10^6
Glass (flint)	10	30×10^6
Quartz	5	30×10^6
Mica	6	20×10^7
Water(d)*	81	n/a

† Correct value is 1.0006.
* Distilled(d) water is unusual – a high dielectric constant with a reasonably good conductivity.

241

A6.2 **Magnetic materials**

Magnetism is a complex subject. There are *five* types of magnetism, namely *diamagnetism*, *paramagnetism*, *ferrimagnetism*, *ferromagnetism*, and *antiferrimagnetism*.

Diamagnetism, paramagnetism, and antiferrimagnetism are very weak effects with no practical significance. Ferrimagnetism is a relatively small effect but appears in materials with high resistivity. It is therefore of interest and is the basis of ferrite rod aerials and the magnetic cores of radio transformers.

Ferromagnetism is the only strong magnetic effect, and *ferromagnetics* are the most useful magnetic materials. The main feature is the relative permeability μ_r. Since the *BH* curve of a magnetic material is non-linear the slope μ varies; Table A6.3 gives *maximum* values of μ_r.

Table A6.3

Material	μ_r
Air	1
Cobalt	2.5×10^2
Nickel	6×10^2
Mild steel	2×10^3
Iron	5×10^3
Mumetal	1×10^5
Supermalloy	1×10^6

Cobalt, nickel, and iron are the only *elements* which are ferromagnetic. The other materials in the table above are alloys of these elements.

Materials which are not ferromagnetic or ferrimagnetic have μ_r values that are approximately unity.

Answers to problems

Chapter 3

P3.1 (i) 10 A; (ii) −10 A;
(iii) Reversed

P3.2 (i) 10 V; (ii) −10 V;
(iii) −10 V; (iv) 10 V

P3.3 (i) 100 W; (ii) −100 W;
(iii) −100 W

P3.4 (b) and (c)

P3.5 −3 V

P3.6 −11 A

P3.7 2 A; 20 W

P3.8 18 μJ

P3.9 50 J

P3.10 $v = 2.5$ mV

P3.11 $v = 2.5 \times 10^4 t^2$

P3.12 It remains stored in the ideal
capacitance indefinitely.

P3.13 The current continues to flow
and the energy is stored in the
ideal inductance indefinitely.

P3.14 (i) 0; (ii) 100 W;
(iii) ∞; (iv) −100 W

P3.15 (i) ∞; (ii) 0; (iii) 100 W;
(iv) −10 W

Chapter 4

P4.1 1 A; 5/8 A; 3/8 A

P4.2 240 V; 235 V; 230 V;
5.1 A

P4.3 53.33 V

P4.4 200 Ω; 500 Ω

P4.5 (i) 20/71 A; (ii) 2/27 A;
(iii) 27/28 A; (iv) 4 A

P4.6 (i) 27/28 A; (ii) 4 A;
(iii) 1 A; (iv) 0.81 A (right to
left); (v) 3.095 A

P4.7 (a) 5 A in parallel with 2 Ω
(b) −2 A in parallel with 3 Ω
(c) 2.5 A in parallel with 2 Ω
(d) ∞ A in parallel with 0 Ω

P4.8 (a) 10 V in series with 2 Ω
(b) −6 V in series with 3 Ω
(c) 12 V in series with 2 Ω
(d) ∞ V in series with ∞ Ω

P4.9 (a) −10 V; 10/3 Ω
(b) 20 V; 17 Ω
(c) 66 V; 10 Ω

P4.10 (a) 1 A; 7 Ω
(b) −0.9 A; 20/9 Ω
(c) 8 A; 5 Ω

P4.11 20/71 A

P4.12 1 A

P4.13 27/28 A

P4.14 0.297 A

P4.15 1 A

P4.16 0.129 A

P4.17 1 Ω

P4.18 0.297 A

P4.19 9.29 V

Chapter 5

P5.1	166.7 W
P5.2	Average power = 0
P5.4	5 A; 5.77 A
P5.5	2.5 V; 2.89 V
P5.6	$0.54Y$; $0.584Y$
P5.7	$70.7\cos(t - 0.7854)$
P5.8	3183.1 Ω; 3183.2 Ω;
	31.4 mA; 89.55° leading
P5.9	60.5°
P5.10	2.006 A; 4.58° lagging;
	400.5 W
P5.11	8.66∠−60° mA; 43.29 mW
	8.66∠60° mA; 43.29 mW
	20∠−30° mA; 173.3 mW
	20∠30° mA; 173.3 mW
P5.12	49.5 mH
P5.13	57.4 μF
P5.14	0.67 μF
P5.15	7118 Hz; 2 A; 4472 V
P5.16	(a) 675; (b) 791; (c) 925
P5.17	(a) 168 Ω; (b) 25 kΩ
P5.18	5 Ω; 79.6 mH
P5.19	150∠0° V; 15.8 ∠ 18.45° A
P5.20	$(4 - j15)$ Ω
P5.21	$(1 - j8)$ Ω
P5.22	17.68 A
P5.23	75.4 ∠ 55.2°
P5.24	75.4 ∠ 55.2°
P5.25	36.7 W; 2.22 W;
	27.8 W; 6.66 W
P5.26	11 W; 9.34 W

Chapter 6

P6.1	(a) 2 A; (b) 0.148 A;
	(c) $(0.148 + 1.85e^{-4.22t})$ A
P6.2	(a) −22 V; (b)−7.3 V
P6.3	−1.16 V
P6.4	3.34 V
P6.5	2.466 V
P6.6	$2.73e^{-303.3t}$ mA;
	$340.58(1 - e^{-303.3t})$ V
P6.7	$1.349e^{-10.64t}\sin(157.76t)$
P6.8	$20e^{-2t} - 10e^{-t}$
P6.9	$-5(1 + 1000t)e^{-1000t}$
P6.10	198.29 V

Chapter 7

P7.1	25.33 mH; 159.2 Ω
P7.3	8.12 mH
P7.4	1559.4 Hz; 1592.2 Hz

Chapter 8

P8.1	0.35 ∠ 28.8°; 1.051 ∠ 118.8°
P8.2	−2.77 ∠ 56.31°
P8.3	2.8 ∠ 22.03°
P8.4	4 Ω
P8.5	2.008 ∠ 58.6°
P8.6	2.975 ∠ 22.8°; 0.941 ∠ 41.2°
P8.7	$(1 + j1.5)$ Ω
P8.8	29.2 ∠ 49.74°; $(3 + j36.3)$ Ω
P8.9	$(2 + j6.5)$ Ω; $(5 + j5)$ V

Chapter 9

P9.1	1.05×10^6 N along Pb
P9.2	0.707 μC
P9.3	3.73
P9.4	14.14 cm
P9.5	2.225×10^6 V/m at 235°
P9.7	$\sigma/2\varepsilon_0$
P9.8	0.102 m and 0.36 m above P
P9.9	500 V, independent of path
	taken
P9.11	1.798 J
P9.12	(a) 9×10^7 N; (b) 8×10^9 V;
	(c) -9×10^8 J
P9.13	Circle centre $(0, -1.25)$, radius
	0.75
P9.14	$\sigma/2\varepsilon_0$
P9.15	$Q/\pi r\varepsilon_0$, $-3Q/2\pi R\varepsilon_0$;
	4.796×10^6 V/m
P9.16	1.13×10^7 V/m; 1130 V,
	0.885 μF
P9.17	$4\pi\varepsilon_0 ab/(a - b)$
P9.18	5.9 nF; 5.9 μC, 667 kV/m,
	7.87 J/m^3
P9.19	3.79 nF, (a) 428 kV/m,
	143 kV/m; (b) 474 μJ
P9.20	5
P9.21	(a) 1770 μJ/cm^2;
	(b) 885 mN/cm^2

P9.22   (a) 1.13 J;   (b) 0.68 J;
(c) 1.13 J (per cm^2)

P9.23   9.48 cm

P9.24   6.25 cm

P9.25   (a) $(\lambda e/2\pi m\varepsilon_0)^{\frac{1}{2}}$;
(b) 1.26×10^7 m/s;
(c) helical path

Chapter 10

P10.1   4.33 N out

P10.2 $v = 1500$ m/s; E in $-$ve z
direction, B in $-$ve x direction

P10.3   0.0072 Nm

P10.4   15.7 A to the right

P10.5   1.253 cm, 17.3 cm

P10.6 $\mu_{r(max)} = 4800$

P10.7 $\mu_0 Ir/2\pi R^2$, $\mu_0 I/2\pi r$

P10.8 $I\{\ln[(x+b)/(x-b)]\}/4\pi b$

P10.9 1.07×10^{-4} T, 6.4×10^{-5} T

P10.10 6.38×10^{-5} T down

P10.12   (a) 5×10^{-4} T;
(b) 8.11×10^{-5} Wb

P10.13 4.94×10^{-4} T

P10.14 4.64×10^{-6} T

P10.15 1.81×10^{-4} T

P10.16   1.49 A

P10.17   (a) 1.34 A;   (b) 0.49 T

P10.18   6.5 A

Chapter 11

P11.1   25 V

P11.2   31.4 mV

P11.3   31.3 mV

P11.4   10 mV

P11.5   52.13 μV

P11.6   graph (max e.m.f. is ± 4 mV)

P11.7   1.6 μW

P11.8   (a) 80 μH;   (b) 1.44 H

P11.9 $\mu_0 Inl/[(4R^2 + l^2)^{\frac{1}{2}}]$

P11.10 $\mu_0 n^2 l^2 \pi R^2/[(4R^2 + l^2)^{\frac{1}{2}}]$;
146.7 μH;   157.9 μH

P11.11 $\mu_0\{\ln[(D-a)/a]\}/\pi$

P11.12 $\mu_0 I^2/16\pi$

P11.13 $\mu_0 n(b-a)\ln(b/a)/2\pi$;   14.4 μH

P11.14 $(\mu_0 i^2 \ln(b/a))/4\pi$;
$(\mu_0 \ln(b/a))/2\pi$;
0.125 μJ, 0.25 μH/m

Bibliography

Circuits

T.F. Bogart *Electric Circuits* 2nd edn, Glencoe (Macmillan/McGraw-Hill), 1992
J.A. Edminister *Electric Circuits* 2nd edn, Schaum (McGraw-Hill), 1983
J.L. Keown *PSpice and Circuit Analysis* Merrill, 1991
J.W. Nilsson *Electric Circuits* 4th edn, Addison-Wesley, 1993
K.F. Sander *Electric Circuit Analysis* Addison-Wesley, 1992
T.W. Thorpe *Computerised Circuit Analysis with Spice* Wiley, 1992
P.W. Tuinenga *Circuit Simulation and Analysis using PSpice* Prentice Hall, 1992
A. Vladimirescu *The Spice Book* Wiley, 1994

Fields

R.G. Carter *Electromagnetism for Electronic Engineers*, 2nd edn, Chapman and
  Hall, 1992
D. Halliday, R. Resnick, and K.S. Krane *Physics Part 2*, 4th edn, Wiley, 1992
P. Hammond *Electromagnetism for Engineers*, 3rd edn, Pergamon Press, 1986
W.H. Hayt *Engineering Electromagnetics*, 5th edn, McGraw-Hill, 1989
J.D. Kraus and K.R. Carver *Electromagnetics*, 2nd edn, McGraw-Hill, 1982
R.G. Powell *Electromagnetism (programmed text)* Macmillan Education Ltd,
  1990

Index